Before You Turn the Page, Take the **WY-TOPP Online Practice Test**

For Students & Parents

1. Scan the QR Code or Visit:
   lumoslearning.com/book

2. Enter the Access Code: QFV7

3. Start the test

For Educators

1. Scan the QR Code or Visit: lumoslearning.com/book
2. Enter the Access Code: DQV1
3. Get Started

Wyoming Test of Proficiency and Progress (WY-TOPP) Test Prep: 4th Grade Math Practice Workbook and Full-length Online Assessments: WY-TOPP Study Guide

Contributing Author - Jessica Fisher
Contributing Author - Wilma Muhammad
Executive Producer - Mukunda Krishnaswamy
Program Director - Anirudh Agarwal
Designer and Illustrator - Nagendra K V .

Updated in May 2026

ISBN 13: 978-1966084747

Printed in the United States of America

FOR SCHOOL EDITION AND PERMISSIONS, CONTACT US

 **LUMOS INFORMATION SERVICES, LLC**

 PO Box 1575, Piscataway, NJ 08855-1575
www.LumosLearning.com

 Email: support@lumoslearning.com
 Tel: (732) 384-0146
Fax: (866) 283-6471

Lumos Learning
Step Up Your Skills

How to Use the Lumos Program?

Congratulations on choosing the Lumos WY-TOPP Program, a blended program that combines online practice with paper-and-pen skill building.

How to Use the Lumos WY-TOPP Program:

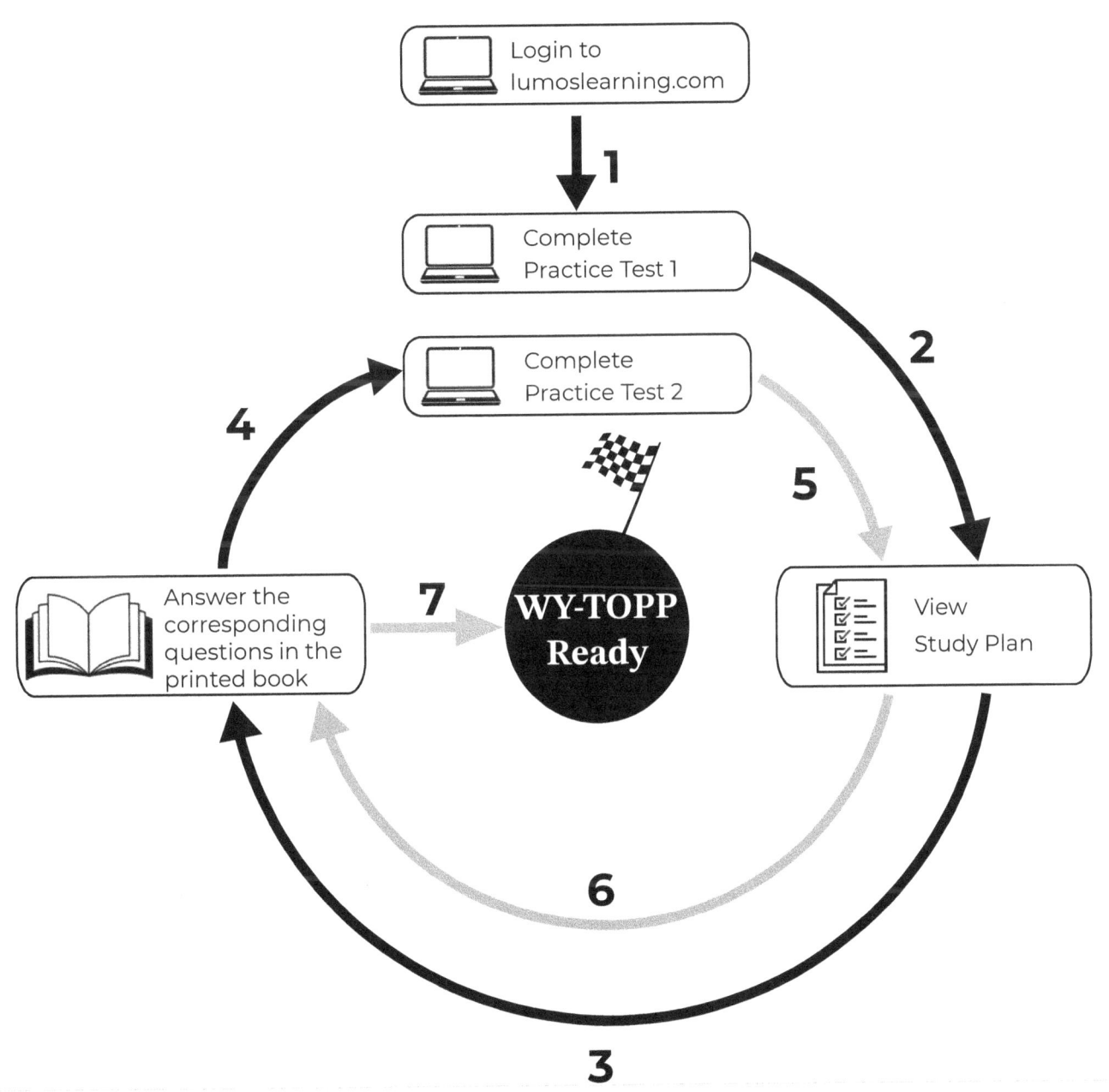

Table of Contents

LumosLearning.com

Chapter 1
Operations and Algebraic Thinking

Lesson 1: Number Sentences

1. Andrew is twice as old as his brother, Josh. Which equation could be used to figure out Andrew's age if Josh's age, n, is unknown?

   Ⓐ $a = n + 2$
   Ⓑ $a = n \div 2$
   Ⓒ $n = a + 2$
   Ⓓ $a = 2 \times n$

2. Mandy bought 28 marbles. She wants to give the same number of marbles to each of her four friends. What equation or number sentence would she use to find the number of marbles each friend will get?

   Ⓐ $28 - 4 = n$
   Ⓑ $28 \div 4 = n$
   Ⓒ $28 + 4 = n$
   Ⓓ $28 \times 4 = n$

3. What number does n represent?
   $3 + 6 + n = 22$

   Ⓐ $n = 9$
   Ⓑ $n = 13$
   Ⓒ $n = 18$
   Ⓓ $n = 31$

4. Cindy's mother baked cookies for the school bake sale. Monday she baked 4 dozen cookies. Tuesday she baked 3 dozen cookies. Wednesday she baked 4 dozen cookies. After she finished baking Thursday afternoon, she took 15 dozen cookies to the bake sale. Which equation shows how to determine the number of cookies that she baked on Thursday?

   Ⓐ $4 + 3 + 4 + n = 15$
   Ⓑ $4 + 3 + 4 = n$
   Ⓒ $4 \times 3 \times 4 \times n = 15$
   Ⓓ $15 \div 11 = n$

5. There are 9 students in Mrs. Whitten's class. She gave each student the same number of popsicle sticks. There were 47 popsicle sticks in her bag. To decide how many sticks each student received, Larry wrote the following number sentence: 47 ÷ 9 = n. How many popsicle sticks were left in the bag after dividing them evenly among the 9 students?

Ⓐ 0
Ⓑ 2
Ⓒ 3
Ⓓ 4

6. Sixty-three students visited the science exhibit. The remainder of the visitors were adults. One hundred forty-seven people visited the science exhibit in all.
How would you determine how many of the visitors were adults?

Ⓐ 63 + 147 = n
Ⓑ 147 ÷ 63 = n
Ⓒ 147 ÷ n = 63
Ⓓ 63 + n = 147

7. Donald bought a rope that was 89 feet long. To divide his rope into 11 foot long sections, he solved the following problem: 89 ÷ 11 = n. How many feet of rope was left over?

Ⓐ 0 feet
Ⓑ 1 foot
Ⓒ 2 feet
Ⓓ 3 feet

8. If 976 - n = 325 is true, Which of the following equations does NOT correctly represent this statement?

Ⓐ 976 + 325 = n
Ⓑ 976 - 325 = n
Ⓒ n + 325 = 976
Ⓓ 325 + n = 976

9. Mary has $54. Jack has n times as much money as Mary does. The total amount of money Jack has is $486. What is n?

Ⓐ 19
Ⓑ 29
Ⓒ 9
Ⓓ None of these

10. Mrs. Williams went to Toys R' US to purchase the following items for each of her 3 children: one bicycle for $150, one bicycle helmet for $8, one arts and crafts set for $34 and one box of washable markers for $2 for each child. What is the total amount she spent before taxes?

    Ⓐ $194.00
    Ⓑ $582.00
    Ⓒ $572.00
    Ⓓ $482.00

11. Write an equation to show how many crayons are below.

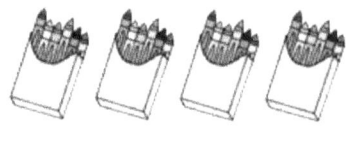

___ × ___ = ___ crayons

12. Alice has 5 bags with 8 pens in each. Which of the following choices represents a number sentence for this situation. Note that more than one option may be correct. Select all the correct answers.

    Ⓐ 8 + 8 + 8 + 8 + 8 = 40
    Ⓑ 5 x 8 = 40
    Ⓒ 5 + 8 = 13
    Ⓓ 8 x 8 = 64

13. Create an equation from the following situation: Tim had a box of chocolates. He started with 18 chocolates, but then gave 6 to his friends. How many does he have left?

14. John draws a regular hexagon. Each side measures 12 centimeters. He also draws a rhombus. The perimeter of the hexagon and the rhombus are the same. How long was each side of the rhombus? Shade the cells to indicate the correct answer.
Note: Each shaded cell is equivalent to 2 cm.

15. Jose purchased 4 books and 8 pens. Each book costs $3, and each pen costs $5. If he gave $100 to the shopkeeper, how much change did he receive back? Circle the correct answer.

   Ⓐ $52
   Ⓑ $48
   Ⓒ $62
   Ⓓ $38

CHAPTER 1 → Lesson 2: Real World Problems

1. There are four boxes of pears. Each box has 24 pears. How many pears are there in total?

   Ⓐ 72 pears
   Ⓑ 48 pears
   Ⓒ 96 pears
   Ⓓ 88 pears

2. Trevor has a collection of 450 baseball cards. He wants to place them into an album. He can fit 15 baseball cards on each page. How can Trevor figure out how many pages he will need to fit into an album all of his cards?

   Ⓐ By adding 450 and 15
   Ⓑ By subtracting 15 from 450
   Ⓒ By multiplying 450 by 15
   Ⓓ By dividing 450 by 15

3. Bow Wow Pet Shop has 12 dogs. Each dog had 4 puppies. How many puppies does the shop have in all?

   Ⓐ 16 puppies
   Ⓑ 12 puppies
   Ⓒ 48 puppies
   Ⓓ 36 puppies

4. Markers are sold in packs of 18 and 24. Yolanda bought five packs of 18 markers and ten packs of 24 markers. How many markers did she buy altogether?

   Ⓐ 42 markers
   Ⓑ 432 markers
   Ⓒ 320 markers
   Ⓓ 330 markers

5. Each box of cookies contains 48 cookies. About how many cookies would be in 18 boxes?

   Ⓐ 100 cookies
   Ⓑ 500 cookies
   Ⓒ 1,000 cookies
   Ⓓ 2,000 cookies

6. At RTA Elementary School, there are 16 more female teachers than male teachers. If there are 60 female teachers, how can you find the number of male teachers in the school?

   Ⓐ Subtract 16 from 60
   Ⓑ Multiply 16 by 60
   Ⓒ Add 16 to 60
   Ⓓ Divide 60 by 16

7. Mrs. Willis wants to purchase enough Christmas ornaments so that every student can decorate 3 each. She has 24 students in her class. How many Christmas ornaments does she need to buy?

   Ⓐ 56
   Ⓑ 72
   Ⓒ 62
   Ⓓ 58

8. Jane needs to sell 69 cookie boxes for her scout troop. She has already sold 36 cookie boxes. How many more cookie boxes does she need to sell ?

   Ⓐ She needs to sell 36 more boxes.
   Ⓑ She needs to sell 39 more boxes.
   Ⓒ She needs to sell 33 more boxes.
   Ⓓ She needs to sell 69 more boxes.

9. RTA Elementary School had a 3-day food drive. On Monday, the school collected 10 soup cans and 16 cereal boxes. On Tuesday, the school collected 23 soup cans and 32 cereal boxes. On Wednesday, the school collected 36 soup cans and 44 cereal boxes. How many cereal boxes did the school collect in all?

   Ⓐ 76 cereal boxes
   Ⓑ 92 cereal boxes
   Ⓒ 161 cereal boxes
   Ⓓ 69 cereal boxes

10. There are 546 students in Hope School and 782 students in Trent School. How many more students are in Trent School than in Hope School?

   Ⓐ 136 students
   Ⓑ 236 students
   Ⓒ 144 students
   Ⓓ 244 students

11. Luke has two bags of pennies. The first bag has 8 pennies and the second bag has 6 times as many pennies as the first bag. How many pennies are in the second bag? Match each number to the correct name by darkening the circle.

	Multiplier	Number in original set	Number in second bag
48	○	○	○
8	○	○	○
6	○	○	○

12. Lisa invented a machine that would triple the number of pencils. A group of friends went to try it out, each having a different number of pencils to start.
Fill in the table below using the rule of tripling the number of pencils

Friends	Starting # of pencils	Multiplier	Expression	Final # of pencils
Lee	2	3		
Timmy		3	3x3	
Katie	4		4x3	
Julia	6	3	6x3	
Isabel		3		27

13. Tommy's fish tank had 8 gallons of water. He wanted to clean it so he drained 7 gallons. Once Tommy was finished, he added 9 times the amount that was in the tank. Shade the cells to show the total amount of water in the tank.
Note: One shaded cell is equivalent to 2 gallons of water in the fish tank.

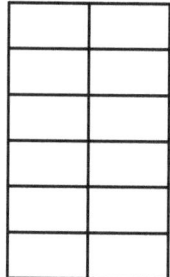

14. Charlie has $938. He wants to buy gifts for his 15 family members. He wants to spend the same amount of money (in dollars) for each person. With the remaining money, he buys a pen worth $3. How much money will he have left over after his purchases? Write your answer in the box below.

$ ⟨_____⟩

CHAPTER 1 →Lesson 3: Multi-Step Problems

1. Kristian purchased four textbooks which cost $34.99 each, and a backpack that cost $19.98. Estimate the total cost of the items he purchased. (You do not need to consider tax.)

   Ⓐ $90.00
   Ⓑ $160.00
   Ⓒ $120.00
   Ⓓ $55.00

2. During the last three games of the season, the attendance at the Tigers' home games was 14,667; 16,992; and 18,124. Estimate the total attendance for these three games. Round to the nearest thousand.

   Ⓐ 60,000
   Ⓑ 45,000
   Ⓒ 50,000
   Ⓓ 47,000

3. Steven keeps his baseball cards in an album. He has filled 147 pages of the album. He can fit 9 cards on each page. Which of the following statements is true?

   Ⓐ Steven has more than 2,000 baseball cards.
   Ⓑ Steven has between 1,000 and 1,500 baseball cards.
   Ⓒ Steven has between 1,500 and 2,000 baseball cards.
   Ⓓ Steven has less than 1,000 baseball cards.

4. Jam jars can be packed in large boxes of 60 or small boxes of 25. There are 700 jam jars to be shipped. The supplier wants to use the least number of boxes possible, but the boxes cannot be only partially filled. How many large boxes will the supplier end up using?

   Ⓐ 10 large boxes
   Ⓑ 11 large boxes
   Ⓒ 12 large boxes
   Ⓓ It is not possible to ship all 700 jars.

5. Allison needs 400 feet of rope to put a border around her yard. She can buy the rope in lengths of 36 feet. How many 36 foot long ropes will she need to buy?

   Ⓐ 9 ropes
   Ⓑ 10 ropes
   Ⓒ 11 ropes
   Ⓓ 12 ropes.

6. Katie and her friend went to the county fair. They each brought a $20.00 bill. The admission fee was $4.00 per person. Ride tickets cost 50 cents each. If each ride required two tickets per person, how many rides was each girl able to go on?

   Ⓐ 16 rides
   Ⓑ 32 rides
   Ⓒ 8 rides
   Ⓓ 20 rides

7. The population of the Bahamas is 276,208. The population of Barbados is 263,584. Which of the following statements is true of the total population of these two places?

   Ⓐ It is less than 500,000.
   Ⓑ It is between 500,000 and 550,000.
   Ⓒ It is between 550,000 and 600,000.
   Ⓓ It is more than 600,000.

8. Corey hopes to have 500 rocks in his collection by his next birthday. So far he has two boxes with 75 rocks each, a box with 85 rocks, and two boxes with 65 rocks each. How many more rocks does Corey need to gather to meet his goal?

   Ⓐ 145 rocks
   Ⓑ 135 rocks
   Ⓒ 275 rocks
   Ⓓ 235 rocks

9. Keith is helping his grandmother roll quarters to take to the bank. Each roll can hold 40 quarters. Keith's grandmother has told Keith that he can keep any leftover quarters, once the rolling is done. If there are 942 quarters to be rolled, how much money will Keith get to keep?

   Ⓐ $4.50
   Ⓑ $5.50
   Ⓒ $5.75
   Ⓓ $3.00

10. Assuming you are working with whole numbers, which of the following is not possible?

   Ⓐ Two numbers have a sum of 18 and a product of 72.
   Ⓑ Two numbers have a sum of 25 and a product of 100.
   Ⓒ Two numbers have a sum of 19 and a product of 96.
   Ⓓ Two numbers have a sum of 25 and a product of 144.

11. George and Michael are both in fourth grade, but attend different schools. George goes to Hillside Elementary and Michael goes to Sunnyside Elementary. Hillside has 7 fourth grade classes with 18 students in each class. Sunnyside has 5 fourth grade classes with 21 students in each class. Mark which of the following are correct responses.

Ⓐ Hillside has 126 fourth grade students.
Ⓑ Sunnyside has 100 fourth grade students.
Ⓒ Sunnyside has fewer fourth grade students than Hillside.
Ⓓ There are 21 more students at Hillside than Sunnyside.
Ⓔ There are 26 more students at Hillside than Sunnyside.

12. Charlie sells paintings. He charges 35 dollars for a large painting and m dollars for a small painting. He sold 8 large paintings and 6 small paintings and earned $412. How much did he charge for each small painting? Write your answer in the box below.

13. Solve each of the problems and match it with the correct answer.

	$9	$20	$10
Karen spent $122 on Christmas presents at the mall and $118 buying presents online. She gave presents to twelve of his family members. She spent the same amount of money on each of them. How much money did she spend for each member of the family?	◯	◯	◯
Jose purchased 12 books and 9 pens. Each book costs $6, and each pen costs $2. If he gave $100 to the shopkeeper, how much change did he receive back?	◯	◯	◯
John bought 25 pens. Each pen costs $8. He sold 15 pens at the rate of $10 per pen. At what rate (selling price per pen) did he sell the rest of them, if he made a profit of $40?	◯	◯	◯

CHAPTER 1 →Lesson 4: Number Theory

1. Andrew has a chart containing the numbers 1 through 100. He is going to put an "X" on all of the multiples of 10 and a circle around all of the multiples of 4. How many numbers will have an "X", but will not be circled?

Ⓐ 3
Ⓑ 4
Ⓒ 5
Ⓓ 8

2. Which number is a multiple of 30?

Ⓐ 3
Ⓑ 6
Ⓒ 60
Ⓓ 50

3. Use the Venn diagram below to respond to the following question.

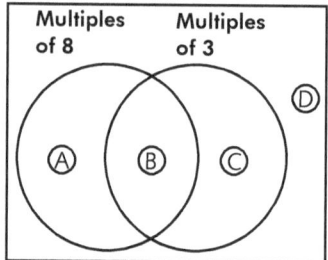

In which region of the diagram would the number 72 be found?

Ⓐ Region A
Ⓑ Region B
Ⓒ Region C
Ⓓ Region D

4. Which number can divide 28 evenly?

Ⓐ 3
Ⓑ 6
Ⓒ 7
Ⓓ 5

5. Use the Venn diagram below to respond to the following question. Which of the following numbers would be found in Region D?

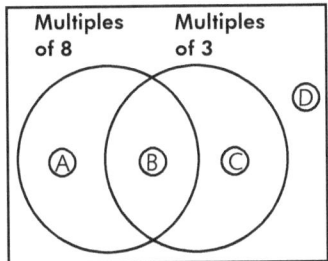

Ⓐ 41
Ⓑ 53
Ⓒ 62
Ⓓ All of the above

6. Which of these sets contains no composite numbers?

Ⓐ 97, 71, 59, 29
Ⓑ 256, 155, 75, 15
Ⓒ 5, 23, 87, 91
Ⓓ 2, 11, 19, 51

7. Choose the set that consists of only prime numbers.

Ⓐ 2, 4, 8, 12
Ⓑ 13, 15, 17, 19
Ⓒ 2, 5, 23, 29
Ⓓ 3, 17, 29, 81

8. Which of the following sets include factors of 44?

Ⓐ 2, 4, 11, 22
Ⓑ 0, 2, 6, 11
Ⓒ 2, 4, 12, 22
Ⓓ 4, 8, 12, 22

9. Which number completes the following number sentences?
$72 \div \underline{\quad} = 6$
$\underline{\quad} \times 6 = 72$

Ⓐ 6
Ⓑ 8
Ⓒ 12
Ⓓ 16

10. If Carla wants to complete exactly 63 push ups during her workout, it is best if she does her push-ups in sets of _____.

Ⓐ 13
Ⓑ 8
Ⓒ 7
Ⓓ 17

11. Read each number below and mark whether it is a factor of 24, 30, or both.

	Factor of 24	Factor of 30	Factor of 30 and 24
2	○	○	○
3	○	○	○
5	○	○	○
8	○	○	○
10	○	○	○
12	○	○	○

12. Identify the prime number and write it in the box given below:

13, 15, 9, 100, 28, 77

13. Circle all the factors of 24

1	2	3	4
5	6	7	8
9	10	11	12
13	14	15	16
17	18	19	20
21	22	23	24

14. Charlie has a chart containing the numbers 1 through 100. He is going to put an "X" on all of the multiples of 6 and a circle around all of the multiples of 8. Which of the following statements are correct? Choose all the correct answers.

Ⓐ 12 numbers are circled.
Ⓑ 9 numbers are circled but do not have X on them.
Ⓒ 4 numbers are circled and also have X on them.
Ⓓ 12 numbers have X on them but are not circled.

CHAPTER 1 →Lesson 5: Patterns

1. **Which of the following is NOT a true statement about the number sequence below?**
   26, 39, 52, 65, 78, 91

   (A) The numbers are decreasing by 13.
   (B) The sequence contains both even and odd numbers.
   (C) The numbers are increasing by 13.
   (D) The numbers are all multiples of 13.

2. **What would be the next three numbers in this pattern?**
   14, 21, 28, 35,

   (A) 42, 50, 58
   (B) 42, 49, 56
   (C) 42, 49, 58
   (D) 42, 48, 54

3. **Which of the following is NOT true of this pattern?**
   125,000; 150,000; 175,000; 200,000; 225,000

   (A) The numbers are not descending.
   (B) The numbers are all even.
   (C) The numbers are increasing by 25,000.
   (D) The numbers are all multiples of 50,000.

4. **What is the missing number in this pattern?**
   24, 36, ___, 60, 72

   (A) 50
   (B) 54
   (C) 48
   (D) 46

5. **Study the following pattern. Then find the next two terms.**
   0, 10, 8, 18, 16, 26, 24, ___, ___

   (A) 36, 34
   (B) 34, 44
   (C) 32, 30
   (D) 34, 32

6. Darren was skip-counting by 5's starting from 2. He said, "2, 7, 12, 17, . . ." After a while, he noticed a pattern in the numbers. Based on the pattern, which of the following numbers will Darren eventually say?

    Ⓐ 720
    Ⓑ 275
    Ⓒ 187
    Ⓓ 271

7. Assume this pattern continues. Which number would not be part of the sequence?
  12, 24, 36, 48, 60, . . .

    Ⓐ 112
    Ⓑ 108
    Ⓒ 120
    Ⓓ 96

8. What is the rule in the following pattern?
  8, 12, 16, 20, 24, 28

    Ⓐ add 4
    Ⓑ subtract 8
    Ⓒ add 3
    Ⓓ add 6

9. Warren likes to eat spoonfuls of dried strawberries out of his bowl of cereal. His bowl has 20 of them. He can fit 5 strawberries on the spoon at one time. How many are still left in the bowl after the third spoonful?

    Ⓐ 15 strawberries
    Ⓑ 5 strawberries
    Ⓒ 10 strawberries
    Ⓓ 1 strawberry

10. Larry produced exactly 12 clown wigs in an 8 hour day, for 3 days. Which pattern shows how many clown wigs were made in 5 weeks?

    Ⓐ 3, 8, 12; 3, 8, 12; 3, 8, 12; 3, 8, 12
    Ⓑ 3, 8, 12; 3, 8, 13; 3, 8, 14; 3, 8, 15; 3, 8, 16
    Ⓒ 3, 8, 12; 3, 8, 24; 3, 8, 36; 3, 8, 48; 3, 8, 60
    Ⓓ 3, 8, 12; 3, 8, 144; 3, 8, 1728; 3, 8, 17,726

11. Select the shape that would come next in each pattern.

	▭	□	○	△
▭ ○ □ ▭	○	○	○	○
△ ○ △ ○ △ ○	○	○	○	○
□ △ ○ ▭ □ ○	○	○	○	○
○ ▭ □ △ ○ ▭	○	○	○	○

12. Fill in the missing boxes in the table to complete each pattern or sequence of numbers Start from the left side of the table and fill in the missing boxes to complete the sequence.

1		3	4		6
2	4		8	10	
	12	18		30	36
10		18	22	26	

13. Find the rule for this IN-OUT table. Circle the correct answer choice.

IN	OUT
72	8
63	7
45	5

Ⓐ Divide by 8
Ⓑ Subtract by 56
Ⓒ Subtract by 64
Ⓓ Divide by 9

14. Determine the rule for this table. Then use the rule to answer the following question. What number will come out when the IN value is 25? Enter your answer in the blank box.

IN	OUT
12	8
15	11
20	16
25	

15. Assume this pattern continues. Which numbers would be part of the sequence? Note that more than one option may be correct.
84, 91, 98, 105 .....
Select all the correct answers.

Ⓐ 343
Ⓑ 288
Ⓒ 245
Ⓓ 167

End of Operations and Algebraic Thinking

Chapter 1
Operations and Algebraic Thinking
Lesson 1: Number Sentences

Question No.	Answer	Detailed Explanation
1	D	It requires multiplication to find out the amount for twice as many. The symbol for multiplication is x. If **n** represents Josh's age, then **a** represents Andrew's age.
2	B	Mandy is making 4 equal groups out of 28. Therefore, 28 divided by 4 equal the number of marbles each friend receives.
3	B	To find n, we need to get it alone by subtracting the other numbers. This is an equation that needs to stay balanced, so what is done on one side of the = sign must be done on the other side. If we subtract 9 (6+3) from both sides, we have n = 13.
4	A	It is known that Cindy's mother baked 4 + 3 + 4 dozens of cookies plus an unknown number (n). The correct equation adds the amount baked Monday through Wednesday and adds the unknown (n).
5	B	47 divided by 9 = 5 with a remainder of 2.
6	D	There is a difference between the number of visitors to the science exhibit and the number of adult visitors. Subtract 63 from 147 to find n. The inverse equation is the correct answer: 63 + n = 147
7	B	89 divided by 11 is 8 with a remainder of 1. The remainder is the number of feet left over.
8	A	Adding 976 and 325 is the opposite of what the problem is stating: what number **subtracted** from 976 = 325.
9	C	Divide 486 by 54. 486 ÷ 54 = 9. Jack has 9 times as much money as Mary does.

Question No.	Answer	Detailed Explanation
10	B	For each child, Mrs. Williams spent $150 + 8 + 34 + 2 = $194.00. However, the beginning of the problem states she is shopping for all three of her children so you will need to determine her full total. For three children, she would spend a total of $194.00 x 3 = $582.00.
11	4x6=24	Since there are 4 boxes, with 6 crayons in each box, to find the total number of crayons, multiply 4 and 6 together, which equals 24.
12	A & B	Each of the 5 bags have 8 pens, so we can either multiply 5 x 8 or add 8 together 5 times (8 + 8 + 8 + 8 + 8) because multiplication is repeated addition.
13	18 - 6 = 12	Let the number of chocolates Tim had be N He gave 6 to his friends. Hence, the balance will be N - 6 So, the number of chocolates left with him will be 18 - 6 = 12
14	18cm	Total No. of Rows x Columns:2 x 6 Cells to be highlighted:9 A regular hexagon has six equal sides. Therefore, perimeter of the hexagon = 6 x 12 = 72 cm. A rhombus has four equal sides. Let the length of each side be s. perimeter of the rhombus = 4 x s = perimeter of the hexagon = 72 cm 4 x s = 72; s = 72 ÷ 4 = 18 cm.
15	B	This is a two-step problem. First, we calculate the total cost of 4 books and 8 pens; Total cost = (4 x 3) + (8 x 5) = 12 + 40 = $52. Next, we subtract the total cost from the amount Jose gave to the shopkeeper to calculate the change he receives back; 100 - 52 = $48.

Lesson 2: Real World Problems

Question No.	Answer	Detailed Explanation
1	C	Each box has 24 pears and there are 4 boxes, so that is 24 x 4, which equals 96.
2	D	The total collection is 450. Each page will have 15 cards or parts of the whole. When you divide 450 by 15, the answer tells how many pages are needed.
3	C	There are 12 dogs in Bow Wow Pet Shop. Since, each dog has 4 puppies, the total number of puppies will be 12 x 4 which is equal to 48.
4	D	This problem requires 2 steps. There were (5) 18 count packs bought. Multiply 5 x 18 = 90. There were (10) 24 count packs bought. Multiply 10 x 24 = 240. Add the two products together to find out the sum or total amount bought. 240 + 90, which equals 330.
5	C	The question begins with the word, "about", which means to estimate. Begin by rounding off the number of cookies. 48 rounds off to 50. Multiply that estimate by the number of boxes: 50 x 18 = 900. Of the four estimate choices, 1,000 is the closest estimate. Also note 18 can be rounded to 20.
6	A	This is a subtraction problem. There are 60 female teachers, however, there are 16 more of them than male teachers. Hence, the number of male teachers will be 60 - 16.
7	B	Each student will decorate 3 ornaments. There are 24 students. That is 24, 3 times or 24 x 3, which equals 72.
8	C	This problem is asking the difference between how many cookies need to be sold and how many are already sold. Therefore, 69 - 36=33.
9	B	The question only asks for the total number of cereal boxes. Everything pertaining to soup cans is extra, unnecessary information. Add 16 + 32 + 44.
10	B	To find the difference between how many more students are at Trent School than Hope school, subtract the amounts: 782 - 546.

Question No.	Answer	Detailed Explanation

| 11 | | |

	Multiplier	Number in original set	Number in second bag
48	○	○	●
8	○	●	○
6	●	○	○

He had 8 pennies in the first bag, so that is the number in the original set. It says the second bag has 6 times as many pennies as the first bag, so 6 is our multiplier. 8 x 6 = 48 pennies in the second bag.

12					
		Starting # of pencil	Multiplier	Expression	Final # of pencil
	Lee	2	3	**2×3**	**6**
	Timmy	**3**	3	3x3	**9**
	Katie	4	**3**	4x3	**12**
	Julia	6	3	6x3	**18**
	Isabel	**9**	3	**9×3**	27

| 13 | 10 | Total No. of Rows x Columns:6 x 2
Cells to be highlighted:5

To find the number of gallons of water in the tank, we need to use the words in the situation. He started with 8 and drained 7 , so 8 – 7 = 1. 9 times 1 is 9 so he then added 9 gallons, so 1 + 9 = 10. |

| 14 | $5 | This is a two step problem. First, we divide the amount Charlie has ($938) by the number of his family members (15). This is a problem on division, in which the remainder has to be interpreted.

15) 938 (62
    90
    38
    30
     8
The remainder is $8. So, Charlie is left with $8 after buying gifts for his family members.
Next, we subtract the cost of the pen ($3) from $8 to find the amount Charlie has after his purchases; 8 - 3 = $5. |

Lesson 3: Multi-Step Problems

Question No.	Answer	Detailed Explanation
1	B	Round off the cost of the textbooks to the nearest dollar. $34.99 is close to $35.00. Round off the price of the backpack from $19.98 to $20.00. The estimated amount Kristian paid can be found by calculating 4 x $35.00 + $20.00. 4 x $35.00 = $140.00; $140.00 + 20.00 = $160.00
2	C	To find the estimated total, round each number to the nearest thousand, then add. 14,667 rounds to 15,000; 16,992 rounds to 17,000; and 18,124 rounds to 18,000. 15,000 + 17,000 + 18,000 = 50,000
3	B	To decide which choice is true, use estimation. Steven has filled about 150 pages in his album. (147 is close to 150.) There are about 10 cards on each page. Therefore, he has about 1,500 cards. (150 x 10 = 1,500) Since both of the numbers were rounded up, this estimate is an overestimate. Steven, therefore, has slightly less than 1,500 cards. The second choice is the most reasonable.
4	A	The best option for the supplier would be to use as many large boxes as possible. The supplier would be able to ship 600 jars in 10 large boxes and 100 jars in 4 small boxes. The supplier could not use any more large boxes, because 11 large boxes would contain 660 jars. That would leave 40 more jars to be shipped. That cannot be done without sending a partially-filled small box.
5	D	If Allison were to buy 10 ropes, she would have 360 feet of rope. (36 x 10 = 360) Buying one more rope would bring her total up to 396 feet. (360 + 36 = 396) She is still 4 feet short. Therefore, Allison will have to buy 12 ropes in order to get at least 400 feet of rope.
6	A	Because the problem is asking about the number of rides **each** girl could go on, you can focus just on Katie. Katie brought $20.00 with her to the fair. After paying the $4.00 admission fee, she has $16.00 left over. If each ride required two tickets, and the tickets are worth 50 cents each, then each ride costs $1.00 per person. With $16.00 left over, Katie could go on 16 rides.
7	B	An estimated sum could be found by using compatible numbers. 276,208 is close to 275,000; 263,584 is close to 265,000. 275,000 + 265,000 = 540,000, which is between 500,000 and 550,000.

Question No.	Answer	Detailed Explanation
8	B	First, total up the rocks that Corey already has in his collection: (2 x 75) + (2 x 65) + 85 = 150 + 130 + 85 = 365. Then subtract what he has from 500: 500 - 365 = 135.
9	B	Divide the total number of quarters by 40 to see how many rolls can be made. The remainder will be the quarters that Keith gets to keep. 942 ÷ 40 = 23 Remainder 22. There can be 23 rolls, with 22 quarters left over. Keith will keep the 22 quarters. The value of 22 quarters (4 quarters make a dollar) is $5.50.
10	C	To solve this problem, first, express the products, namely, 72, 100, 96, 144 in terms of their prime factors (simplest factors). Then by trial and error, try to make two numbers from them (by multiplication), such that their sums are 18, 25, 19, 25. In the following, product of underlined numbers is the first number, and product of other numbers is the second number. Thus, 72 = <u>2 x 2 x 3</u> x 3 x 2. This can be grouped as 12 x 6 and 12 + 6 = 18. 100 = <u>2 x 2 x 5</u> x 5. This can be grouped as 20 x 5 and 20 + 5 = 25. 96 = 2 x 2 x 2 x 2 x 2 x 3 . This cannot be grouped into two numbers whose sum is 19. 144 = <u>2 x 2 x 2 x 2</u> x 3 x 3. This can be grouped as 16 x 9 and 16 + 9 = 25. Solution for choice A: 12 and 6 Solution for choice B: 20 and 5 Solution for choice D: 16 and 9 There is no solution for choice C.
11	A, C & D	To find the total number of students at each school, multiply the number of classes times the number of students in each. Hillside has 126 students and Sunny-side has 105 students. Therefore A is true, B is false, and C is true. To figure out how many more students Hillside has, subtract 126 – 105 = 21 students, so D is true.
12	$22	Amount Charlie earned by selling 8 large paintings = 8 x 35 = $280. We subtract this amount from his total earnings ($412) to find the selling price of 6 small paintings; 412 - 280 = $132. Selling price of 1 small painting = m; Selling price of 6 small paintings = 6 x m = 132. Therefore, m = 132 ÷ 6 = $22

Question No.	Answer	Detailed Explanation

13

	$9	$20	$10
Karen spent $122 on Christmas presents at the mall and $118 buying presents online. She gave presents to twelve of his family members. She spent the same amount of money on each of them. How much money did she spend for each member of the family?		◯	
Jose purchased 12 books and 9 pens. Each book costs $6, and each pen costs $2. If he gave $100 to the shopkeeper, how much change did he receive back?			◯
John bought 25 pens. Each pen costs $8. He sold 15 pens at the rate of $10 per pen. At what rate (selling price per pen) did he sell the rest of them, if he made a profit of $40?	◯		

(1) First, we find the total amount Karen spent; 122 + 118 = $240. Next, we find the amount of money Karen spent for each member of the family (12 members) by dividing the total amount ($240) by 12; 240 ÷ 12 = $20.

(2) This is a two-step problem. First, we calculate the total cost of 12 books and 9 pens; Total cost = (12 x 6) + (9 x 2) = 72 + 18 = $90. Next, we subtract the total cost from the amount Jose gave to the shopkeeper to calculate the change he receives back; 100 - 90 = $10.

(3) This is a three-step problem. First, we calculate the amount John invested by multiplying the number of pens by the cost of each pen; 25 x 8 = $200.

John sold 15 pens at the rate of $10 per pen. Selling price of 15 pens at the rate of $10 per pen = 15 x 10 = $150.

Question No.	Answer	Detailed Explanation
		John sold 15 pens at the rate of $10 per pen. Selling price of 15 pens at the rate of $10 per pen = 15 x 10 = $150.

Next, we calculate the selling price of the remaining pens. Let t be the selling price of the remaining 10 pens.

profit = selling price - cost price (Equation 1)

In this problem, profit = $40

cost price = $200

selling price = selling price of 15 pens at the rate of $10 per pen + selling price of the remaining 10 pens = 150 + t

Substituting these values in the Equn. 1, we get,

40 = (150 + t) - 200 = t - 50

t = 40 + 50 = $90 = selling price of the remaining 10 pens

In the third step, we find the selling price of 1 pen (which is in the group remaining 10 pens). Let the selling price of 1 pen = s. Then, we have s x 10 = 90; s = 90 ÷ 10 = $9.

Therefore, John sold the remaining 10 pens at the rate of $9 per pen. |

Lesson 4: Number Theory

Question No.	Answer	Detailed Explanation
1	C	Multiples are the products of two numbers. Skip count, recite timetables or refer to a chart to find all of the multiples of both numbers, from 1 to 100. Listing these multiples may also be helpful: 10 is 10, 20, 30, 40, 50, 60, 70, etc. 4 is 4, 8, 12, 16, 20, 24, 28, 32, etc. All of the multiples of 10 will have an x and all multiples of 4 will be circled. The question to the problem is **how many numbers will have an x, but** *not* **be circled.** In the list of 10 multiples, circle all the multiples of 4. Count the number of multiples of 10 that do not have a circle.
2	C	The multiple of 30 must be a product of 30 x another number. 60 = 30 x 2.
3	B	In this diagram, Section A would contain the multiples of 8 which are not multiples of 3. Section C would contain the multiples of 3 which are not multiples of 8. Section B would contain multiples of both 8 and 3. Section D would list numbers that are **not** multiples of 8 or 3. 8 x 9 = 72 and 3 x 24 = 72, so 72 would be found in Section B.
4	C	Using the inverse relationship between multiplication and division to choose the number that is a factor of 28. 28 is a multiple of 7, so 7 is a factor of 28.
5	D	Refer to the lists of the multiples of 8 and 3. The D section of the Venn diagram is for numbers that are not multiples of 8 or 3. 41, 53, and 62 are not multiples of 3 or 8.
6	A	Any whole number greater than 1 is either classified as composite or prime. Therefore, this question is asking you to find the set that contains only prime numbers. A prime number is a number that has only 2 factors - itself and 1. The first set of numbers fits this criteria. All four of the numbers in the set have only two factors.
7	C	Prime numbers are numbers greater than 1 that have only two factors: themselves and the number 1.
8	A	Choose the set that gives the product of 44 when multiplied with another number in that set.
9	C	12 x 6 = 72, so 72 ÷ 12 = 6
10	C	63 is a multiple of 7. 63 is not a multiple of 13, 8, or 17. In order to do exactly 63 push-ups, Carla is best to do sets of 7.

11.

	Factor of 24	Factor of 30	Factor of 30 and 24
2			○
3			○
5		○	
8	○		
10		○	
12	○		

To be a factor, the number has to go into the number evenly. Looking at 2 and 3, we see both 24 and 30 are divisible, so the answer for both 2 and 3 is factor of both. 5 and 10 both go evenly into 30 but not 24, while 8 and 12 both go evenly into 24 but not 30.

12	13	A prime number is a number that has only 2 factors - itself and 1. From the list of numbers given, 13 is the only prime number. The correct answer is 13.

13		1, 2, 3, 4, 6, 8, 12, 24 needs to be selected.

Factors are numbers that go evenly into the number. Because 1 x 24 = 24, 2 x 12 = 24, 3 x 8 = 24, and 4 x 6 = 24, 1, 2, 3, 4, 6, 8, 12, and 24 are all factors of 24.

14	A,C&D	Multiples of 8 : There are 10 numbers up to 80 (8, 16, etc.) which are divisible by 8. There are 2 numbers between 81 to 100 which are divisible by 8 (88 and 96). Therefore, there are 12 numbers between 1 and 100 which are divisible by 8. Therefore, option (A) is correct.

Multiples of 6 : There are 10 numbers up to 60 (6, 12, etc.) which are divisible by 6. There are 6 numbers between 61 to 100 which are divisible by 6 (66, 72, 78, 84, 90, 96). So, there are 16 numbers between 1 and 100 which are divisible by 6.

Among the multiples of 8, 4 numbers (24, 48, 72 and 96) are divisible by 6. Therefore, there are 12 - 4 = 8 numbers that are circled but do not have X on them. Therefore, option (B) is wrong.
24, 48, 72 and 96 are divisible by both 6 and 8. Therefore, option (C) is correct.

Among the multiples of 6, 4 numbers (24, 48, 72 and 96) are divisible by 8. Therefore, there are 16 - 4 = 12 numbers which have X on them but are not circled. Therefore, option (D) is correct.

Lesson 5: Patterns

Question No.	Answer	Detailed Explanation
1	A	Identify whether there is adding, subtracting multiplying or dividing to create this pattern. If the numbers increase, it is adding or multiplying; if the numbers decrease, it is subtracting or dividing. The next thing is to determine the relationship between the numbers in the pattern, starting with the first two: 26 and 39 is an increase. Ask yourself: Is the increase from adding or multiplying. If the numbers are multiples of a number or numbers, the increase is from multiplying. Otherwise, it is from adding. How many numbers are between 26 and 39 is the answer to how many numbers are being added. Be sure to read the question carefully.
2	B	The numbers are increasing, so determine if this pattern is adding or multiplying. The numbers in the problem are multiples of 7, so a continued list of the multiples of 7 are needed to continue this pattern.
3	D	Carefully read each answer choice and check the given numbers. Ascending means increasing and descending means decreasing. Even numbers are all numbers that are multiples of 2. Multiples of 50,000 would be any numbers that are products of 50,000 x another whole number. In this problem, the numbers are increasing (not decreasing), even, and they are multiples of 25,000. Therefore, statements made in options (A), (B), and (C) are true. But all the numbers are not multiples of 50,000. For example, 1,25,000 / 50,000 = 2.5. Therefore, statement made in option (D) is not true. We have to choose the statement which is not true. So, option (D) is the correct answer.
4	C	Identify the relationship between the first two and last two numbers. In this case, they are increasing through addition or multiplication. Ask yourself: 24 + ___ = 36 or are all of these numbers multiples of another number. Ask the same question regarding 60 and 72. Using addition, reverse the action: 36 - 24 = ? and 72 - 60 = ?That is the missing number.
5	D	The numbers in this pattern are increasing and decreasing: 0 to 10 is an increase of 10 numbers. 10 to 8 is a decrease of 2 numbers. 8 to 18 is an increase of 10 numbers. Continue to calculate the pattern in this fashion to determine the next two numbers.
6	C	Continue to skip count by 5's from 17. List the numbers to determine which one will be a part of the pattern. So far, all of the numbers end in a 2 and a 7.
7	A	The numbers in this pattern are multiples of 12. Continue counting by 12's or adding 12 to the last number in the pattern, up to 120 in order to determine which answer choices are multiples of 12.

Question No.	Answer	Detailed Explanation
8	A	The rule is add 4 to the number each time.
9	B	Determine how many strawberries are left (subtract) after each spoonful: 20 - 5 = 15; 15 - 5 = 10; 10 - 5 = 5.
10	C	The first number in the pattern is 3, which represents the number of days. Next is 8, which represents the length of the day. The third number is 12, which represents how many wigs were made. This pattern repeats 4 more times, always beginning with the 3 and 8. The third number will be an additional 12, representing 12 more wigs made: 12 + 12 = 24, 24 + 12 = 36, and so on.

11

	▭	□	○	△
▭○□▭			○	
△○△○△○				○
□△○▭□○	○			
○▭□△○▭		○		

To find the next shape in the pattern, look at the shapes on the left and see what comes after the last shape. In the first one, the yellow circle comes after the orange square, so we select the yellow circle.

12

1	**2**	3	4	**5**	6
2	4	**6**	8	10	**12**
6	12	18	**24**	30	36
10	**14**	18	22	26	**30**

To find the pattern, look at the differences of each pattern. In the first, they are adding by 1. In the second pattern, they are adding by 2. In the third pattern, they are adding by 6. In the last pattern, they are adding by 4. Knowing the pattern, all you have to do is add the given pattern to the previous term, or subtract the given pattern from the next term.

Question No.	Answer	Detailed Explanation
13	D	For each number in the IN column, the corresponding number in the OUT column is less. Therefore, check whether you get the number in the OUT column by subtracting or dividing the number in the IN column. Also to be considered is what could be done to 72 to get 8. If we divide 72 by 9, we get 8. Now, check whether this rule "divide by 9" holds for the other two numbers also. 63 ÷ 9 = 7, 45 ÷ 9 = 5. The rule "divide by 9" holds good in all the three cases.
14	21	For each number in the IN column, the corresponding number in the OUT column is less. Therefore, check whether you get the number in the OUT column by subtracting or dividing the number in the IN column. Also to be considered is what could be done to 12 to get 8. If we subtract 4 from 12, we get 8. Now, check whether this rule "subtract 4" holds for the other two numbers also. 15 - 4 = 11, and 20 - 4 = 16. The rule "subtract 4" holds good in all the three cases. So, when the IN value is 25, the OUT value = 25 - 4 = 21.
15	C	Note that all the numbers in the pattern are multiples of 7. 245 and 343 are divisible by 7; 245 ÷ 7 = 35 and 343 ÷ 7 = 49. Therefore, 245 and 343 would be part of the sequence. 167 and 288 are not divisible by 7. Therefore, they would not be part of the sequence.

Chapter 2
Number & Operations in Base Ten

Lesson 1: Place Value

1. What number can be found in the ten-thousands digit of 291,807?

Ⓐ 9
Ⓑ 1
Ⓒ 2
Ⓓ 0

2. Consider the number 890,260.
   The 8 is found in the _____ place.

   Ⓐ ten-thousands
   Ⓑ millions
   Ⓒ thousands
   Ⓓ hundred-thousands

Place Value Chart

Hundred-billions	Ten-billions	Billions	Hundred-millions	Ten-millions	Millions	Hundred-thousands	Ten-thousands	Thousands	Hundreds	Tens	Ones

3. What number correctly completes this statement?
   9 ten thousands = _____ thousands

   Ⓐ 90
   Ⓑ 900
   Ⓒ 9
   Ⓓ 19

4. Which number is in the thousands place in the number 984,923?

   (A) 9
   (B) 8
   (C) 4
   (D) 2

5. What is the value of the 8 in 683,345?

   (A) 80
   (B) 800
   (C) 8,000
   (D) 80,000

6. Which number equals 4 thousands, 6 hundreds, 0 tens, and 5 ones?

   (A) 465
   (B) 4,605
   (C) 4,650
   (D) 4,065

7. What number is in the tens place in 156.25?

   (A) 1
   (B) 5
   (C) 6
   (D) 2

8. Which number equals 2 ten thousands, 1 hundred thousand, and 3 ones

   (A) 120,003
   (B) 210,003
   (C) 102,003
   (D) 213,000

9. Which answer shows the value of each 7 in this number: 7,777?

   (A) 7,000, 700, 70, 7
   (B) 7 x 7 x 7 x 7
   (C) 700,000, 70,000, 700, 70
   (D) 7 + 7 + 7 + 7

10. Mrs. Winters went to the bank with eight 100 dollar bills. She wanted to replace them with all 10 dollar bills. How many 10 dollar bills will the bank give her in exchange?

Ⓐ 800 ten dollar bills
Ⓑ 8,000 ten dollar bills
Ⓒ 8 ten dollar bills
Ⓓ 80 ten dollar bills

11. Select the correct value for each number

	5	50	500
How many hundreds are in 500?	○	○	○
How many tens are in 500?	○	○	○
How many ones are in 500?	○	○	○

12. Select the correct value for each number

	9	90	900
How many hundreds are in 900?	○	○	○
How many tens are in 900?	○	○	○
How many ones are in 900?	○	○	○

13. Which number equals 8 millions, 5 tens? Circle the correct answer

Ⓐ 800,050
Ⓑ 8,000,500
Ⓒ 8,000,005
Ⓓ 8,000,050

14. John has $500. Karen has 10 times as much money. How much money does Karen have? Write your answer in the box below

CHAPTER 2 →Lesson 2: Compare Numbers and Expanded Notation

1. Arrange the following numbers in ascending order (least to greatest).
   62,894; 26,894; 26,849; 62,984

   Ⓐ 62,984; 62,894; 26,894; 26,849
   Ⓑ 26,894; 26,849; 62,984; 62,894
   Ⓒ 26,849; 62,984; 62,894; 26,894
   Ⓓ 26,849; 26,894; 62,894; 62,984

2. Which of the following statements is true?

   Ⓐ 189,624 > 189,898
   Ⓑ 189,624 > 189,246
   Ⓒ 189,624 < 189,264
   Ⓓ 189,624 = 189,462

3. Which number will make this statement true?
   198,888 > _____

   Ⓐ 198,898
   Ⓑ 198,879
   Ⓒ 198,889
   Ⓓ 199,888

4. Which statement is NOT true?

   Ⓐ 798 < 799
   Ⓑ 798 > 789
   Ⓒ 798 < 789
   Ⓓ 798 = 798

5. Write the expanded form of this number.
   954,351

   Ⓐ 90,000 + 5,000 + 400 + 30 + 5 + 1
   Ⓑ 900,000 + 50,000 + 4,000 + 300 + 50 + 1
   Ⓒ 900,000 + 54,000 + 300 + 50 + 1
   Ⓓ 900,000 + 50,000 + 4,000 + 300 + 51

6. What is the standard form of this number?
   30,000 + 200 + 50

   Ⓐ 30,250
   Ⓑ 32,500
   Ⓒ 325,000
   Ⓓ 3,250

7. Write the standard form of:
   8 ten thousands, 4 thousands, 1 hundred, 6 ones

   Ⓐ 84,160
   Ⓑ 84,106
   Ⓒ 8,416
   Ⓓ 84,016

8. Write 1,975,206 in expanded form.

   Ⓐ 1,000,000 + 9,000,000 + 7,000 + 500 + 20 + 6
   Ⓑ 1,000,000 + 9,000,000 + 70,000 + 5,000 + 200 + 60
   Ⓒ 100,000 + 900,000 + 70,000 + 5,000 + 200 + 6
   Ⓓ 1,000,000 + 900,000 + 70,000 + 5,000 + 200 + 6

9. What is 300,000 + 40,000 + 20 + 5 in standard form?

   Ⓐ 34,025
   Ⓑ 3,425
   Ⓒ 340,025
   Ⓓ 342,005

10. Write the standard form for 2 hundred thousands, 1 ten thousand, 4 hundreds, 1 ten, 9 ones.

    Ⓐ 210,419
    Ⓑ 201,419
    Ⓒ 200,419
    Ⓓ 21,419

11. Compare the numbers then select <, >, or = to make the sentence true.

	<	>	=
4,145 ___ 4,451	○	○	○
31,600 __ 63,100	○	○	○
49 _____ 49	○	○	○
831 _____ 381	○	○	○

12. Write the standard form of this number, placing the comma in the correct place.
2,000 + 500 + 9

13. Compare the numbers then select <, >, or = to make the sentence true.

	<	>	=
124 __ 789	○	○	○
12,947 __19,247	○	○	○
412 __ 412	○	○	○
94,811__81,944	○	○	○

14. The number of toothpicks produced each day in a factory is given below.

Toothpicks Produced	
Day	Number of Toothpicks Produced
Monday	825,347
Tuesday	825,374
Wednesday	825,743
Thursday	852,743
Friday	852,743

On which day did the factory produce the least number of toothpicks? Mark the correct answer.

Ⓐ Monday
Ⓑ Tuesday
Ⓒ Wednesday
Ⓓ Thursday

CHAPTER 2 →Lesson 3: Rounding Numbers

1. **Round 4,170,154 to the nearest hundred.**

   Ⓐ 4,200,000
   Ⓑ 4,170,100
   Ⓒ 4,170,000
   Ⓓ 4,170,200

2. **Round 4,170,154 to the nearest thousand.**

   Ⓐ 4,200,000
   Ⓑ 180,000
   Ⓒ 4,170,000
   Ⓓ 4,179,200

3. **Round 4,170,154 to the nearest ten thousand.**

   Ⓐ 4,200,000
   Ⓑ 4,170,000
   Ⓒ 4,179,000
   Ⓓ 4,179,200

4. **Round 4,170,154 to the nearest hundred thousand.**

   Ⓐ 4,200,000
   Ⓑ 4,180,000
   Ⓒ 4,179,000
   Ⓓ 4,100,000

5. **Round 4,170,154 to the nearest million.**

   Ⓐ 4,000,000
   Ⓑ 4,180,000
   Ⓒ 4,179,000
   Ⓓ 5,000,000

6. **Round 424,819 to the nearest ten.**

   Ⓐ 400,820
   Ⓑ 424,810
   Ⓒ 424,020
   Ⓓ 424,820

7. The soda factory bottles 2,451 grape and 3,092 orange sodas each day. About how many of the two types of soda are bottled each day? Round the numbers to the nearest hundred and add them.

   Ⓐ 5,400 sodas
   Ⓑ 5,000 sodas
   Ⓒ 5,600 sodas
   Ⓓ 5,500 sodas

8. The Campbells have 3,000 books in their personal library. If 1,479 of the books are fiction, and the rest are non-fiction. How many are non-fiction? Round to the nearest hundred.

   Ⓐ 2,500 books
   Ⓑ 1,400 books
   Ⓒ 1,500 books
   Ⓓ 1,600 books

9. Solve; Give an estimate as the final answer.
   7000 - 4258 = ?

   Ⓐ 2,600
   Ⓑ 3,000
   Ⓒ 2,000
   Ⓓ 6,858

10. The school has $925 to spend on new books for the library. Each book costs $9.95. Estimate how many books can be bought.

   Ⓐ 70
   Ⓑ 80
   Ⓒ 90
   Ⓓ 120

11. Match each statement with the way in which it is rounded.

	Nearest 10	Nearest 100	Nearest 1,000	Nearest 10,000
4,893 rounded to 4,890	○	○	○	○
15,309 rounded to 20,000	○	○	○	○
32,350 rounded to 32,000	○	○	○	○
523 rounded to 500	○	○	○	○

12. Round 14,623 to the nearest hundred.

( )

13. Match each statement with the way in which it is rounded

	Nearest 10	Nearest 100	Nearest 1,000	Nearest 10,000
19,989 rounded to 19,990	○	○	○	○
86,415 rounded to 90,000	○	○	○	○
909 rounded to 1,000	○	○	○	○
8,525 rounded to 8,500	○	○	○	○

14. If the number 750,025 is rounded to the nearest tfen thousand to get 760,000, What would be the list of possible digits that could go in the thousands place? Circle the correct answer.

Ⓐ 5,6,7,8,9
Ⓑ 6,7,8,9
Ⓒ 0,1,2,3,4
Ⓓ 0,1,2,3,4,5

CHAPTER 2 →Lesson 4: Addition & Subtraction

1. **What number acts as the identity element in addition?**

   Ⓐ -1
   Ⓑ 0
   Ⓒ 1
   Ⓓ None of these

2. **Which of the following number sentences illustrates the Commutative Property of Addition?**

   Ⓐ 3 + 7 = 7 + 3
   Ⓑ 9 + 4 = 10 + 3
   Ⓒ 11 + 0 = 11
   Ⓓ 2 + (3 + 4) = 2 + 7

3. **What number makes this number sentence true?**
   10 + ___ = 10

   Ⓐ 10
   Ⓑ 1
   Ⓒ 100
   Ⓓ 0

4. **Find the sum.**
   24 + 37 + 76 + 13

   Ⓐ 140
   Ⓑ 150
   Ⓒ 151
   Ⓓ none of these

5. **Find the difference.**
   702 - 314 = _____

   Ⓐ 388
   Ⓑ 412
   Ⓒ 312
   Ⓓ 402

6. What is the sum of 0.55 + 6.35?

   Ⓐ 6.09
   Ⓑ 0.85
   Ⓒ 6.90
   Ⓓ none of these

7. Find the difference.
   7.86 - 4.88

   Ⓐ 2.98
   Ⓑ 3.02
   Ⓒ 2.08
   Ⓓ 1.98

8. Find the sum.
   156 + 99 =

   Ⓐ 256
   Ⓑ 245
   Ⓒ 265
   Ⓓ 255

9. There are 1,565 pictures on the disks. Only 1,430 of them are in color. How many pictures are not in color?

   Ⓐ 135 pictures
   Ⓑ 2,995 pictures
   Ⓒ 2,195 pictures
   Ⓓ 995 pictures

10. James got 300 coins while diving in the game, The Amazing World of Gumball Splashmasters! However, he hit 2 birds and lost 12 coins. During the second round, he got 250 coins and hit no birds. How many coins did he have at the end of the second round?

   Ⓐ 500 coins
   Ⓑ 542 coins
   Ⓒ 548 coins
   Ⓓ 538 coins

11. **Select all of the following expressions that will equal 4,189**

    Ⓐ 4,002 + 187
    Ⓑ 6,100 – 2,189
    Ⓒ 12,555 – 8,366
    Ⓓ 859 + 3,985

12. **Fill in the missing digit(s) to make the addition problem true.**

```
   1525
 +2_98
 _____
   3923
```

13. Enter the missing digit(s) to make the subtraction problem true.

```
   7598
  -2_98
 _____
   5500
```

14. **Find the sum: 4,927 + 5,098. Circle the correct answer.**

    Ⓐ 9025
    Ⓑ 9925
    Ⓒ 10,025
    Ⓓ 10,015

CHAPTER 2 →Lesson 5: Multiplication

1. Assume a function table has the rule "Multiply by 6." What would the OUT value be if the IN value was 8?

   Ⓐ 14
   Ⓑ 48
   Ⓒ 16
   Ⓓ 32

2. Solve.
   26 x 8 = ___

   Ⓐ 206
   Ⓑ 168
   Ⓒ 182
   Ⓓ 208

3. The number sentence 4 x 1 = 4 illustrates which mathematical property?

   Ⓐ The Associative Property of Multiplication
   Ⓑ The Identity Property of Multiplication
   Ⓒ The Distributive Property
   Ⓓ Commutative Property of Multiplication

4. Find the product of 17 x 6.

   Ⓐ 102
   Ⓑ 84
   Ⓒ 119
   Ⓓ 153

5. If the IN value is 0, what is the OUT value?

RULE: Divide by 9

IN	OUT
0	??
9	1
81	9
99	11

Ⓐ 2
Ⓑ 0
Ⓒ 9
Ⓓ 1

6. Which of the following number sentences illustrates the Associative Property of Multiplication?

Ⓐ 4 x 0 = 0
Ⓑ 77 x 1 = 1 x 77
Ⓒ (2 x 4) x 5 = 2 x (4 x 5)
Ⓓ 13 x 7 = (10 x 7) + (3 x 7)

7. Solve.
   4 x 3 x 6 = _____

Ⓐ 48
Ⓑ 72
Ⓒ 56
Ⓓ 64

8. Find the exact product of 5 x 20 x 8.

Ⓐ 800
Ⓑ 560
Ⓒ 80
Ⓓ 900

9. Complete the following statement:
   "In a multiplication sentence, if a factor is 1, then _____ "

Ⓐ the other factor is also 1
Ⓑ the product is also 1
Ⓒ the other factor and the product are the same
Ⓓ the product is 0

10. Steve has 7 pages in his stamp collection book. Each page holds 20 stamps. How many total stamps does Steve have in his collection?

Ⓐ 14
Ⓑ 6
Ⓒ 140
Ⓓ 8

11. Select the correct statement for each multiplication fact.

	True	False
75 x 5 = 375	○	○
28 x 12 = 334	○	○
72 x 9 = 648	○	○
45 x 21 = 945	○	○

12. Fill in the missing numbers to complete each multiplication sentence.

5	x		=	65
23	x	14	=	
	x	34	=	340
52	x		=	104

13. Circle the correct answer to the multiplication problem 14 x 12.

Ⓐ 140
Ⓑ 168
Ⓒ 186
Ⓓ 120

14. The van traveled 1,458 miles every day from Monday through Friday. How many miles did it travel in all? Write your answer in the box below.

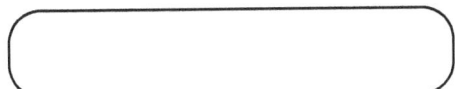

CHAPTER 2 →Lesson 6: Division

1. **What role does the number 75 play in the following equation?**
   300 ÷ 75 = 4

   Ⓐ It is the dividend.
   Ⓑ It is the quotient.
   Ⓒ It is the divisor.
   Ⓓ It is the remainder.

2. **Which of the following division expressions will have no remainder?**

   Ⓐ 73 ÷ 9
   Ⓑ 82 ÷ 6
   Ⓒ 91 ÷ 7
   Ⓓ 39 ÷ 9

3. **Divide these blocks into 2 equal groups. How many will be in each group? (Blocks are able to be broken)**
   Note: 1 flat = 10 rods. 1 rod = 10 cubes

   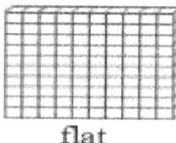 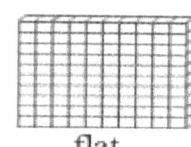 

       flat   flat   flat   rods   cubes

   Ⓐ 352
   Ⓑ 176
   Ⓒ 152
   Ⓓ 132

4. **Fill in the blank:**
   480 ÷ 6 = (420 ÷ 6) + (60 ÷ ____)

   Ⓐ 400
   Ⓑ 6
   Ⓒ 80
   Ⓓ 480

5. **What amount would be in each group if this number were divided into 6 groups?**
   Note: 1 flat = 10 rods. 1 rod = 10 cubes

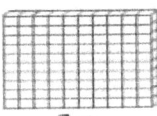

flat

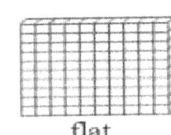

flat

rod

Ⓐ 35
Ⓑ 45
Ⓒ 25
Ⓓ 15

6. **Find the quotient:**
   694 ÷ 2 = ____

   Ⓐ 342
   Ⓑ 327
   Ⓒ 347
   Ⓓ 332

7. **What is the remainder when 100 is divided by 12?**

   Ⓐ 4
   Ⓑ 6
   Ⓒ 8
   Ⓓ 2

8. **28 x 6 = 168**
   Find a related fact to the one shown above.

   Ⓐ 28 ÷ 16 = 8
   Ⓑ 168 ÷ 6 = 28
   Ⓒ 168 ÷ 28 = 8
   Ⓓ 168 ÷ 8 = 16

9. **Thomas family took 96 bottles of spring water to the family reunion picnic. If they had purchased 4 identical cases, how many bottles were in each case?**

   Ⓐ 12
   Ⓑ 48
   Ⓒ 24
   Ⓓ 18

10. Each of Mrs. Harris' 7 children collected the same number of exotic insects for their individual displays. How many insects did each child have for show and tell, if they had a total of 112 insects?

Ⓐ 112
Ⓑ 16
Ⓒ 7
Ⓓ 42

11. Select all of the following problems that do not have remainders in their answer.

Ⓐ 28 ÷ 7
Ⓑ 16 ÷ 5
Ⓒ 17 ÷ 3
Ⓓ 42 ÷ 5

12 Fill in the blank that will make the following division sentence true.

512 ÷ _____ = 256

13. Complete the following table.

Dividend	Divisor	Quotient	Remainder
128	8	16	0
435	7		
350	6		

14. John bought 6 packs of beads. Each pack has the same number of beads. Altogether, he has 1,494 beads. How many beads are in each pack? Circle the correct answer.

Ⓐ 259 beads
Ⓑ 249 beads
Ⓒ 247 beads
Ⓓ 229 beads

End of Number and Operations in Base Ten

Chapter 2
Number & Operations in Base Ten
Lesson 1: Place Value

Question No.	Answer	Detailed Explanation
1	A	Place values are read from right to left, beginning with the "ones" place, "tens", "hundreds", "thousands", "ten thousands", "hundred thousands", "millions", etc. If you were to write the number in the boxes below, you see the 9 is in the ten-thousand column.

Place Value Chart

Hundred-billions	Ten-billions	Billions	Hundred-millions	Ten-millions	Millions	Hundred-thousands	Ten-thousands	Thousands	Hundreds	Tens	Ones

Question No.	Answer	Detailed Explanation
2	D	Begin naming the place values for each number from the right. Number 9 is in the ten thousands place. Place values increase by multiplying 10: 1 ten is 10, 10 tens is a hundred, 10 hundreds is a thousand, etc.
3	A	Multiply 9 x 10,000 to find 90,000.
4	C	Number 3 is in the "ones" place. Number 2 is in the "tens" place. Number 9 is the "hundreds" place. The "thousands" place is next.
5	D	The 8 is in the ten thousands place, which is 8 x 10,000.
6	B	Write the 4 in the thousands place, the 6 in the hundreds place, the 0 in the tens place and the 5 in the ones place.w

Question No.	Answer	Detailed Explanation
7	B	Numbers to the right of the decimal point begin with the value of tenths, hundredths, etc. Numbers to the left of the decimal place are the ones, tens, hundreds, etc.

one millions / hundred thousands / ten thousands / one thousands / hundreds / tens / ones / tenths / hundredths / thousandths / ten thousandths / hundred thousandths / millionths

9 , 6 0 5 , 8 7 2 . 1 4 5 6 7 8

Question No.	Answer	Detailed Explanation
8	A	Though not stated as such in the problem, the digit in the hundred thousands place is written first. The 2 ten thousands is written next: 2 ten thousands is 2 x 10,000. The next place that has any value is the ones place, which has 3. The thousands and hundreds place have no value, so zeros are placed there.
9	A	Write the numbers in expanded notation, which shows the entire value of the number written out. 7 in the thousands place is written as 7,000. 7 in the hundreds place is written as 700. 7 in the tens place is written as 70. 7 in the ones place is written as 7.
10	D	There are ten 10 dollar bills in $100. Therefore, there are 80 ten-dollar bills in $800.
11		

	5	50	500
How many hundreds are in 500?	○		
How many tens are in 500?		○	
How many ones are in 500?			○

Question No.	Answer	Detailed Explanation

12

	9	90	900
How many hundreds are in 900?	○		
How many tens are in 900?		○	
How many ones are in 900?			○

13   D   Place values are read from right to left, beginning with the "ones" place, "tens", "hundreds", "thousands", "ten thousands", "hundred thousands", "millions" etc

Place Value Chart

Hundred-billions	Ten-billions	Billions	Hundred-millions	Ten-millions	Millions	Hundred-thousands	Ten-thousands	Thousands	Hundreds	Tens	Ones

14 $5000   Karen has 10 times as much money. It means we have to multiply the money John has by 10; 500 x 10 = $5,000.

Lesson 2: Compare Numbers and Expanded Notation

Question No.	Answer	Detailed Explanation
1	D	Ascending order is from least to greatest value. All of these numbers start in the ten thousands place, however the numbers beginning with a 2 have a lesser value. If the numbers are the same in the thousands place, for example, compare the numbers that are different in the hundreds place. If the digits are the same in the hundreds place, compare the tens.
2	B	6 hundreds, 2 tens and 4 ones is greater than 2 hundreds, 6 tens and 4 ones. The numbers are the same in all places above the ten places so you have to look there.
3	B	8 x 10 is greater than 7 x 10.
4	C	Compare all of the digits in the numbers: 9 x 10 is not < 8 x 10.
5	B	This form requires that every digit be multiplied by the multiple of ten that corresponds to its place value and then written as such. For example, the 9 is in the hundred thousands place, which is 9 x 100,000 and written as 900,000. Place a plus sign (+) between each term in the expanded form.
6	A	There are no thousands and no ones in this number. Place a 0 in those positions.
7	B	The standard form is simply the digits in the number placed into their appropriate places. A zero will be placed in the tens place, since there are no tens listed in the problem.
8	D	The standard form is a 7-digit number, so the expanded form must show numbers in the millions place. There are no tens in this number, so in expanded form, this place value is simply omitted.
9	C	Zeros should be written in the thousands and hundreds places. All other digits will be placed according to their corresponding place values.
10	A	There are no thousands in this number, so a 0 will be placed in the thousands place. All other digits are placed according to their corresponding place values.

Question No.	Answer	Detailed Explanation
11		

	<	>	=
4,145 ___ 4,451	O		
31,600 __ 63,100	O		
49 ____ 49			O
831 ____ 381		O	

We need to compare the numbers and fill in with the correct sign.
< means less than, > means greater than, and = means equals.

For the first two, we see that the first number is smaller than the second, so < would fit. The third line has equal numbers, so we would choose =. Finally, the last set of numbers has the first one being the larger, so we would choose >.

Question No.	Answer	Detailed Explanation
12	2509	Notice that there is a two in the thousands place and a 5 in the hundreds. There is not a number for the 10s, so there is a 0 in the tens place, and finally there is a 9 in the ones place. Putting those numbers in, we have 2,509.
13		

	<	>	=
124 __ 789	O		
12,947 __ 19,247	O		
412 __ 412			O
94,811 __ 81,944		O	

Question No.	Answer	Detailed Explanation
14	A	To compare numbers, we start comparing the digits in the highest place value. If they are same, we compare the digits in the next lower place value and so on. In this problem, all the numbers have the same digit (8) in the hundred thousands place. The first three numbers, 825,347, 825,374 and 825,743 have 2 in the ten thousands place. The last two numbers, 852,743 and 852,473 have 5 in the ten thousands place. Therefore, the first three numbers are less than the last two numbers. Next, we have to find the least number among the three numbers; 825,347, 825,374 and 825,743. 825,347 and 825,374 have 3 in the hundreds place. 825,743 has 7 in the hundreds place. Therefore, 825,347 and 825,374 are less than 825,743. Of the two numbers, 825,347 and 825,374, 825,347 is less than 825,374 because the former has 4 in the tens place and the latter has 7 in the tens place. Therefore, the factory produced the least number of toothpicks (825,347) on Monday.

Lesson 3: Rounding Numbers

Question No.	Answer	Detailed Explanation
1	D	Since this number is to be rounded to the nearest hundred, the number to the right of the hundreds place (the tens place) will determine if the hundreds digit is to be rounded up to the next number or stay the same. If the number in the tens place is 0-4, keep the hundreds place the same. If the number in the tens place is between 5 and 9 (inclusive), round the hundred up to the next number. Example 678 would round up to 700; whereas 628 would round down to 600.
2	C	The number in the hundreds place is a 1, so the thousands place will not change. It will remain a zero. The hundreds, tens, and ones places will also become zeros.
3	B	Use the digit in the thousands place to decide how to round. In this number, the thousands place has a zero, so the ten thousands place will remain unchanged when rounding.
4	A	Since the digit in the ten thousands place is 5 or greater (it is a 7), round up to the next hundred thousand.
5	A	Round this number based on the digit that is in the hundred thousands place. If it is a 0-4, the digit in the millions place will remain the same. If the digit in the hundred thousands place is a 5-9, the digit in the millions place will round up to the next digit.
6	D	The digit 9 in the ones place will cause the number to round up to the next ten. When rounding to the nearest ten, the digit in the ones place always becomes a zero.
7	C	First round the two values given in the problem to the nearest hundred. 2,451 rounds to 2,500. 3,092 rounds to 3,100. Then add the rounded values to find the estimated sum. 2,500 + 3,100 = 5,600
8	C	First round each of the values given in the problem to the nearest hundred. 3,000 rounds to 3,000. 1,479 rounds to 1,500 Then subtract the rounded values to find the estimated difference. 3,000 - 1,500 = 1,500
9	B	4258, when rounded off gives 4000. 7000-4000 = 3000. Hence, Option B is the correct answer.
10	C	Using reasoning to estimate the amount of books that could be purchased: Each book costs about $10.00. The library has around $900.00 to spend. Dividing 900 by 10, yields a quotient of 90. The school can purchase about 90 books for its library.

Question No.	Answer	Detailed Explanation

11

	Nearest 10	Nearest 100	Nearest 1,000	Nearest 10,000
4,893 rounded to 4,890	O			
15,309 rounded to 20,000				O
32,350 rounded to 32,000			O	
523 rounded to 500		O		

When rounding to the nearest 10, look at the digit in the ones place. If that digit is 5 or more, add 1 to the digit in the tens place. If the digit in the ones place is 4 or smaller, keep the digit in the tens place as it is. That is the same process for rounding to the nearest 100, 1,000, or 10,000. The only difference is you look at the digit to the right of the digit you want to round. So row 1 was rounded to the nearest 10, row 2 was rounded to the nearest 10,000, row 3 was rounded to the nearest 1,000, and row 4 was rounded to the nearest 100.

Keeping these rules in mind, we can look at the problem in another way. We have to do checking in the ascending order. i.e. from nearest to 10, next nearest to 100 and so on.

(1) 4893 is rounded to 4890. Clearly it is rounded to nearest ten.

(2) 15,309 is rounded to 20,000. Since all the digits in ones place, tens place, hundreds place are zeros (after rounding), we have to check whether 15,309 is rounded to nearest thousand or ten thousand. If it were rounded to nearest thousand, it would have been 15,000 NOT 20,000. So, 15,309 is rounded to nearest ten thousand.

Question No.	Answer	Detailed Explanation
		(3) 32,350 is rounded to 32,000. Since the digits in ones place and tens place are zeros (after rounding), we have to check whether 32,350 is rounded to nearest hundred or thousand. If it were rounded to nearest hundred, it would have been 32,400 NOT 32,000. So, 32,350 is rounded to nearest thousand. (4) 523 is rounded to 500. It is not rounded to nearest ten. If it were, it would have been 520 NOT 500. 523 is rounded to nearest hundred.
12	14,600	To round to the nearest hundred, look at the digit in tens place. Since 2 is below 5, the 6 in the hundreds place stays and the tens and ones place changes to 0.
13		(see table below)

	Nearest 10	Nearest 100	Nearest 1,000	Nearest 10,000
19,989 rounded to 19,990	O			
86,415 rounded to 90,000				O
909 rounded to 1,000			O	
8,525 rounded to 8,500		O		

When rounding to the nearest 10, look at the digit in the ones place. If that digit is 5 or more, add 1 to the digit in the tens place. If the digit in the ones place is 4 or smaller, keep the number in the tens place as it is. That is the same process for rounding to the nearest 100, 1,000, or 10,000. The only difference is you look at the digit to the right of the digit you want to round. So row 1 was rounded to the nearest 10, row 2 was rounded to the nearest 10,000, row 3 was rounded to the nearest 1,000, and row 4 was rounded to the nearest 100..

Keeping these rules in mind, we can look at the problem in another way. We have to do checking in the ascending order. i.e. from nearest to 10, next nearest to 100 and so on.

(1) 19,989 is rounded to 19,990. Clearly it is rounded to nearest ten.
(2) 86,415 is rounded to 90,000. Since all the digits in ones place, tens place, hundreds place are zeros (after rounding), we have to check whether 86,415 is rounded to nearest thousand or ten thousand.

Question No.	Answer	Detailed Explanation
		If it were rounded to nearest thousand, it would have been 86,000 NOT 90,000. So, 86,415 is rounded to nearest ten thousand. (3) 909 is rounded to 1,000. Since the digits in ones place, tens place are zeros (after rounding), we have to check whether 909 is rounded to nearest hundred or thousand. If it were rounded to nearest hundred, it would have been 900 NOT 1,000. So, 909 is rounded to nearest thousand. (4) 8,525 is rounded to 8,500. If 8,525 is rounded to nearest ten, it would have been 8,530 NOT 8,500. Therefore, it is clear that 8,525 is rounded to nearest hundred.
14	A	The given number is between 750,000 and 760,000. When rounding is done, it is rounded to 760,000. Therefore, the number is rounded up. It means the number has to be more than 755,000. Therefore, the digit in the thousands place has to be 5 or more than 5. Therefore, option (A) is correct.

Lesson 4: Addition & Subtraction

Question No.	Answer	Detailed Explanation
1	B	Any number plus 0 is always that number, so 0 is the additive identity.
2	A	The Commutative Property means that changing the order of the addends (the two numbers being added together) does not change the sum.
3	D	The correct answer is D. $10 + 0 = 10$
4	B	The same rules apply when adding more than two numbers: 24 37 76 13
5	A	Line numbers up. The first one is written down and the second number is written underneath. Subtract the number on the bottom from the number on the top. If the number on the top is smaller than the one on the bottom, regroup: you cannot subtract 4 from 2, so take one from the tens place. Oops! That is a zero, so take 1 from the hundreds place. That leaves a 6 in the hundreds place. Bring that 1 to the tens place and then borrow it again. Take it over to the ones place where it is needed. That leaves a 9 in the tens place and the 2 becomes a 12. Now subtract: 702 <u>314</u>
6	C	Numbers with decimal points are added just like whole numbers, except the decimal points must be in alignment directly underneath each other, including the answer: 0.55 <u>6.35</u>
7	A	After aligning the two decimal numbers, subtract as if there is no decimal point. Problem will be similar to subtracting 488 from 786. Eight cannot be taken from 6, so regroup in the tens place. That 8 becomes a 7 and the 6 in the ones place becomes a 16. Now subtract, regrouping again in the hundreds place so that 8 can be subtracted from 17. The 7 in the hundreds place becomes a 6. Plot a decimal point in the answer directly beneath the other decimal points.
8	D	You can round 99 to 100, and then add 156. Then subtract 1.

Question No.	Answer	Detailed Explanation
9	A	Subtract the number of pictures that are color from the total number of pictures: 1,565 - 1,430. Regrouping is not necessary in this problem.
10	D	Total coins at the end of the second game = number of coins won in the first game - coins lost (for hitting the bird) + number of coins won in the second game = 300 - 12 + 250 = 538.
11	A & C	Options A and C are correct. After you add or subtract each expression, you see options A and C equal 4,189.
12	3	Working right to left, we add 5 + 8 = 13 so we regroup the 10 so the 20 becomes 30. 30 + 90 = 120, so we have to regroup the hundreds. 600 + ? = 900. We know that the answer is 300, so the blank is a 3.
13	0	Working right to left, the ones column subtracts to 0, the tens column subtracts to 0, and we have a 5 in the hundreds place, so we are subtracting 0 from the hundreds column.
14	C	$\begin{array}{r} 4927 \\ +5098 \\ \hline 10{,}025 \end{array}$

Lesson 5: Multiplication

Question No.	Answer	Detailed Explanation
1	B	Apply the rule to each IN value: 8 x 6
2	D	This problem could be solved using the Distributive Property: 26 x 8 = (20 + 6) x 8 = (20 x 8) + (6 x 8) = 160 + 48 = 208
3	B	Any number times 1 will always equal that number.
4	A	This problem could be solved using the Distributive Property: 17 x 6 = (10 + 7) x 6 = (10 x 6) + (7 x 6) = 60 + 42 = 102
5	B	The rule in this table is to divide the IN number by 9. 0 ÷ 9 = 0.
6	C	The order in which numbers are grouped and multiplied does not change the product.
7	B	Even without parentheses, the order in which the numbers are multiplied will not change the product: Multiply 4 x 3. Then multiply that product x 6. 4 x 3 = 12 and 12 x 6 = 72.
8	A	This can be solved using mental math. Multiply 5 x 20 = 100. Then multiply 100 x 8 = 800.
9	C	Any number multiplied by 1 will always have that number as the product, which is the Identity Property of Multiplication.
10	C	One page holds 20 stamps. Steve has 7 pages, therefore total stamps with Steve, 7 x 20 = 140.

11

	True	False
75 x 5 = 375	●	○
28 x 12 = 334	○	●
72 x 9 = 648	●	○
45 x 21 = 945	●	○

To find the correct answer, multiply the problem by using your favorite method. The first, third, and fourth rows are true, while the second row is less by 2.

Question No.	Answer	Detailed Explanation

12

5	x	**13**	=	65
23	x	14	=	**322**
10	x	34	=	340
52	x	**2**	=	104

To find the missing number in the first row, figure out what number multiplied by 5 equals 65. That number is 13. The second row is the answer when you multiply 23 x 14. The third and fourth row is like the first row.

13 | B | Using any method you like, multiply 14 x 12 gives us 168.

14. | 7,290 | Monday through Friday is 5 days. Multiply 1,458 by 5 to get the total distance traveled.

Total distance traveled = 1,458 x 5 = (1,000 + 400 + 50 + 8) x 5

= 1,000 x 5 + 400 x 5 + 50 x 5 + 8 x 5

= 5,000 + 2,000 + 250 + 40

= 7,290 miles

Lesson 6: Division

Question No.	Answer	Detailed Explanation
1	C	The number that follows the division symbol ($\div$) is called the divisor.
2	C	$91 \div 7$ will have no remainder, since 91 is a multiple of 7. $7 \times 13 = 91$.
3	B	Count the value of the base 10 blocks that are shown, which is 352. Because there is only one set of blocks, regroup each hundreds flat into 10 tens rods. Regroup each tens rod into ten ones cubes. Place 1 tens rod and 1 ones cube into 2 groups until all are placed. Count the value of the tens blocks shown in both groups. The numbers should be the same.
4	B	Both numbers that make up the dividend (480) must be divided by the divisor (6).
5	A	The number being modeled is 210. Regroup all of the hundreds flats into tens rods. Place 1 tens rod and 1 ones cube into 6 groups until all are placed, regrouping tens for ones when necessary. Count by tens and ones to determine the value amount of blocks in each group. That number should be the same in each group and is the answer.
6	C	Begin by dividing 6 by 2, following these steps in this order: 1. Divide 2. Multiply 3. Subtract 4. Bring down the next number in the dividend. Multiply the 2 by 3, which is the number of times 2 divides into 6. Place the 3 above the 6. $2 \times 3 = 6$. Place that product under the 6 in the dividend. Subtract the two 6's. Bring down the next number in the dividend, which is 9 and write it next to the 0. Repeat the four steps for each number in the dividend.
7	A	$12 \times 8 = 96$, so $100 \div 12$ would have a remainder of 4, since 100 is 4 greater than 96.
8	B	The product in the multiplication problem is the dividend in the division problem. One factor becomes the divisor and the other one becomes the quotient. Therefore, $168 \div 6 = 28$ is related to $28 \times 6 = 168$.
9	C	Divide 96 by 4. $96 \div 4 = (100 - 4) \div 4 = (100 \div 4) - (4 \div 4) = 25 - 1 = 24$
10	B	Divide 112 by 7. $112 \div 7 = (70 + 42) \div 7 = (70 \div 7) + (42 \div 7) = 10 + 6 = 16$

Question No.	Answer	Detailed Explanation
11	A	Once you have divided each of the choices, you see that (A) and (D) are the two options that do not have remainders. The other two do.
12	2	To solve, we need to figure out what number is being divided from 512. We can use guess and check. First off, we know that 512 is getting smaller, so we know it is not being divided by 1. So we try 2. 512 ÷ 2 = 256, so we have found our answer.

13

Dividend	Divisor	Quotient	Remainder
128	8	16	0
435	7	62	1
350	6	58	2

14   B   Divide 1,494 by 6 to find the number of beads in each pack.

```
6) 1494 ( 249
   12
   29
   24
    54
    54
     0
```

Number of beads in each pack = 249 beads

Chapter 3
Number & Operations - Fractions

Lesson 1: Equivalent Fractions

1. **What fraction of these shapes are squares?**

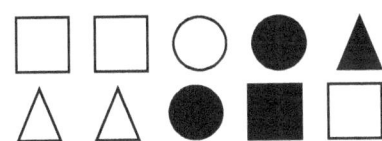

Ⓐ $\frac{1}{4}$ Ⓑ $\frac{4}{6}$

Ⓒ $\frac{4}{10}$ Ⓓ $\frac{1}{3}$

2. **What fraction of these shapes are not circles?**

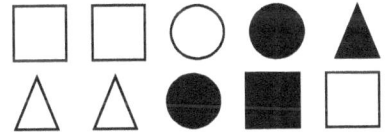

Ⓐ $\frac{3}{7}$ Ⓑ $\frac{8}{10}$

Ⓒ $\frac{7}{10}$ Ⓓ $\frac{1}{3}$

3. **What fraction of the squares are shaded?**

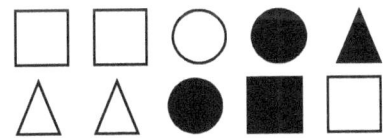

Ⓐ $\frac{1}{4}$ Ⓑ $\frac{1}{10}$

Ⓒ $\frac{1}{3}$ Ⓓ $\frac{3}{4}$

4. What fraction of the shaded shapes are circles?

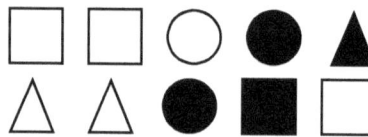

Ⓐ $\frac{2}{10}$ Ⓑ $\frac{1}{3}$

Ⓒ $\frac{2}{2}$ Ⓓ $\frac{2}{4}$

5. Continue the pattern of equivalent fractions:

$$\frac{1}{2}, \frac{2}{4}, \frac{3}{6}, \frac{4}{8} \cdots$$

What fraction would come next in the pattern?

Ⓐ $\frac{1}{3}$ Ⓑ $\frac{1}{16}$

Ⓒ $\frac{5}{10}$ Ⓓ $\frac{3}{4}$

6. Which pair of addends has the fraction $\frac{11}{12}$ as their sum?

Ⓐ $\frac{9}{6} + \frac{2}{6}$ Ⓑ $\frac{7}{12} + \frac{4}{12}$

Ⓒ $\frac{9}{12} + \frac{1}{12}$ Ⓓ $\frac{11}{12} + \frac{1}{1}$

7. Which fraction is equivalent to this model?

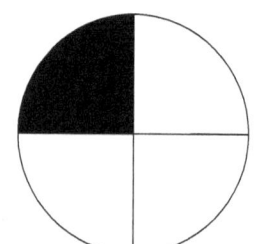

Ⓐ $\frac{1}{5}$ Ⓑ $\frac{3}{7}$

Ⓒ $\frac{2}{7}$ Ⓓ $\frac{4}{16}$

8. Which fraction is equivalent to 8/18?

Ⓐ $\frac{1}{5}$ Ⓑ $\frac{3}{7}$

Ⓒ $\frac{2}{7}$ Ⓓ $\frac{4}{9}$

9. Continue the pattern of equivalent fractions:

$\frac{5}{6}, \frac{10}{12}, \frac{15}{18} \cdots$

What fraction would come next in the pattern?

Ⓐ $\frac{7}{14}$ Ⓑ $\frac{20}{24}$

Ⓒ $\frac{9}{45}$ Ⓓ $\frac{12}{36}$

10. Reduce the fraction $\frac{21}{49}$ to its lowest terms:

Ⓐ $\frac{1}{5}$ Ⓑ $\frac{3}{7}$

Ⓒ $\frac{2}{7}$ Ⓓ $\frac{4}{9}$

11. Reduce the fraction $\frac{44}{99}$ to its lowest terms:

Ⓐ $\frac{1}{5}$ Ⓑ $\frac{3}{7}$

Ⓒ $\frac{2}{7}$ Ⓓ $\frac{4}{9}$

12. Patrick climbed $\frac{4}{5}$ of the way up the trunk of a tree. Jacob climbed $\frac{80}{100}$ of the way up the same tree. To accomplish the same distance as Patrick and Jacob, how far up that tree trunk will Devon have to climb?

Ⓐ $\frac{15}{20}$ Ⓑ $\frac{60}{75}$

Ⓒ $\frac{100}{200}$ Ⓓ $\frac{28}{42}$

13. The cheerleaders ate $\frac{9}{18}$ of a sheet cake. Write this fraction in lowest terms.

    Ⓐ $\frac{1}{9}$ Ⓑ $\frac{1}{2}$

    Ⓒ $\frac{2}{3}$ Ⓓ $\frac{3}{6}$

14. Which group of fractions can all be reduced to $\frac{2}{9}$?

    Ⓐ $\frac{23}{27}$, $\frac{4}{36}$, $\frac{30}{270}$

    Ⓑ $\frac{25}{50}$, $\frac{30}{60}$, $\frac{50}{100}$

    Ⓒ $\frac{4}{18}$, $\frac{6}{27}$, $\frac{50}{225}$

    Ⓓ $\frac{6}{21}$, $\frac{20}{70}$, $\frac{36}{84}$

15. What do these fractions have in common?

$\frac{10}{16}$; $\frac{15}{24}$; $\frac{20}{32}$; $\frac{25}{40}$; $\frac{30}{48}$

    Ⓐ These fractions are equivalent to $\frac{5}{9}$.

    Ⓑ These fractions are equivalent to $\frac{5}{8}$

    Ⓒ These fractions are equivalent to $\frac{10}{12}$.

    Ⓓ These fractions are equivalent to $\frac{4}{8}$.

16. Select whether the fraction pair is equivalent or not equivalent.

	Equivalent	Not Equivalent
$\frac{12}{15}$ and $\frac{3}{5}$	◯	◯
$\frac{18}{24}$ and $\frac{9}{12}$	◯	◯
$\frac{18}{200}$ and $\frac{9}{100}$	◯	◯
$\frac{3}{15}$ and $\frac{3}{25}$	◯	◯

17. Write the simplest form of $\frac{120}{150}$. Write the answer in the box given below.

18. Circle on all of the fractions that can be simplified to $\frac{1}{2}$

Ⓐ $\frac{24}{26}$ Ⓑ $\frac{2}{4}$ Ⓒ $\frac{5}{11}$

Ⓓ $\frac{35}{70}$ Ⓔ $\frac{9}{20}$ Ⓕ $\frac{7}{14}$

19. Which group of fractions are equivalent to $\frac{4}{12}$? Select all the correct answers.

Ⓐ $\frac{1}{3}$, $\frac{2}{5}$, $\frac{3}{9}$

Ⓑ $\frac{1}{3}$, $\frac{2}{6}$, $\frac{3}{9}$

Ⓒ $\frac{1}{3}$, $\frac{2}{5}$, $\frac{5}{20}$

Ⓓ $\frac{6}{18}$, $\frac{12}{36}$, $\frac{15}{45}$

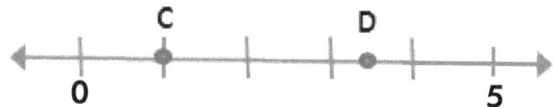

CHAPTER 3 →Lesson 2: Compare Fractions

1. **Where is Point D located on this number line?**

C   D

0   5

   Ⓐ 2.5
   Ⓑ 2
   Ⓒ 3.5
   Ⓓ 3

2. **Which statement is true?**

   Ⓐ $\frac{4}{14} = \frac{6}{21} = \frac{8}{28}$

   Ⓑ $\frac{4}{14} > \frac{6}{21} > \frac{8}{28}$

   Ⓒ $\frac{4}{14} < \frac{6}{21} < \frac{8}{28}$

   Ⓓ $\frac{4}{14} < \frac{6}{21} > \frac{8}{28}$

3. **Compare the two fractions using < = or >:**
   $\frac{3}{12}$ _____ $\frac{3}{18}$
   Ⓐ =
   Ⓑ <
   Ⓒ >

4. **Compare the two fractions using < = or >:**
   $\frac{4}{28}$ _____ $\frac{4}{20}$
   Ⓐ =
   Ⓑ <
   Ⓒ >

5. **Which symbol makes this statement true?**
   $\frac{4}{9} + \frac{3}{9}$ _____ $\frac{6}{9}$
   Ⓐ >
   Ⓑ =
   Ⓒ <

6. Which symbol makes this statement true?

$$\frac{75}{100} - \frac{32}{100} \underline{\quad} \frac{42}{100}$$

Ⓐ >

Ⓑ =

Ⓒ <

7. Which fraction below has a greater value than the fraction being shown?

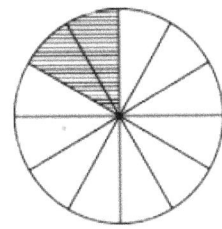

Ⓐ $\frac{2}{20}$

Ⓑ $\frac{2}{15}$

Ⓒ $\frac{2}{10}$

Ⓓ $\frac{2}{100}$

8. A popsicle was left out in the hot sun. As the popsicle melted, which of the following lists of fractions represent the amount of the popsicle remaining after 2 minutes, 4 minutes, and 6 minutes had passed?

Ⓐ $\frac{1}{3}, \frac{1}{2}, \frac{3}{4}$

Ⓑ $\frac{3}{4}, \frac{1}{2}, \frac{1}{3}$

Ⓒ $\frac{1}{3}, \frac{3}{4}, \frac{1}{2}$

Ⓓ $\frac{3}{4}, \frac{1}{3}, \frac{1}{2}$

9. Arrange these models in order from greatest to least:

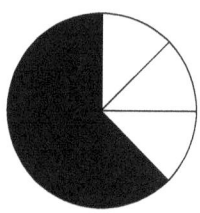

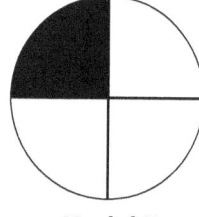

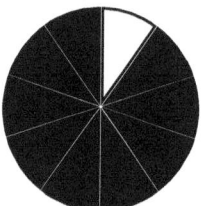

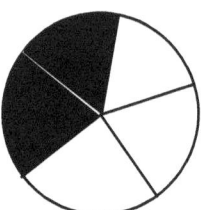

Model A	Model B	Model C	Model D

Ⓐ A, B, C, D
Ⓑ C, A, D, B
Ⓒ C, D, B, A
Ⓓ A, C, D, B

10. Compare the fractions using <, =, or >.

$\frac{692}{1000} + \frac{231}{1000}$ —— $\frac{245}{1000} + \frac{726}{1000}$

Ⓐ <
Ⓑ =
Ⓒ >

11. These fractions are arranged from least to greatest. Which fraction could go in the blank?

$\frac{9}{25}$, —— , $\frac{16}{25}$, $\frac{21}{25}$

Ⓐ $\frac{13}{25}$

Ⓑ $\frac{6}{25}$

Ⓒ $\frac{18}{25}$

Ⓓ $\frac{23}{25}$

12. Marty eats vegetables $\frac{6}{7}$ days out of the week. Dan eats them $\frac{3}{7}$ days out of the week. How many more days does Marty eat vegetables each week?

Ⓐ 2 days
Ⓑ 1 day
Ⓒ 3 days
Ⓓ 4 days

13. The salesman sold $\frac{1}{3}$ of his inventory during a weekend sale. He had hoped to sell an even higher amount. Which fraction could represent the amount of his inventory he had hoped to sell?

Ⓐ $\frac{1}{4}$

Ⓑ $\frac{1}{8}$

Ⓒ $\frac{1}{2}$

Ⓓ $\frac{1}{6}$

14. One fifth of the tourists went to see the natural waterfall on Monday. Five-fifths of them went to see it on Tuesday, and three-fifths of them went to see it on Wednesday.
List the days in ascending order according to the fraction of visitors who visited the waterfall.

Ⓐ Tuesday, Wednesday, Monday
Ⓑ Monday, Wednesday, Tuesday
Ⓒ Wednesday, Tuesday, Monday
Ⓓ Monday, Tuesday, Wednesday

15. What makes $\frac{4}{9} > \frac{4}{11}$?

Ⓐ If numerators are same, a fraction with lower denominator is greater.
Ⓑ If numerators are same, a fraction with higher denominator is greater.
Ⓒ Ninths are larger than elevenths.
Ⓓ Ninths and elevenths are the same size.

16. Select the correct inequality sign that would go in between the fractions

		<	>	=
$\frac{24}{36}$	$\frac{2}{4}$	◯	◯	◯
$\frac{1}{4}$	$\frac{5}{11}$	◯	◯	◯
$\frac{10}{20}$	$\frac{7}{14}$	◯	◯	◯
$\frac{4}{4}$	$\frac{10}{11}$	◯	◯	◯

17. Circle all the fractions that are greater than $\frac{2}{3}$.

Ⓐ $\frac{30}{36}$

Ⓑ $\frac{3}{4}$

Ⓒ $\frac{5}{10}$

Ⓓ $\frac{30}{70}$

Ⓔ $\frac{9}{14}$

Ⓕ $\frac{7}{9}$

18. Which of the following fractions make this statement true?
More than one answer maybe correct. Select all the correct answers
$\frac{3}{5} > $ _____.

Ⓐ $\frac{4}{5}$

Ⓑ $\frac{3}{10}$

Ⓒ $\frac{8}{15}$

Ⓓ $\frac{12}{20}$

CHAPTER 3 →Lesson 3: Adding and Subtracting Fractions

1. **Find the sum:**

    $\frac{3}{12} + \frac{2}{12} =$

    Ⓐ $\frac{5}{24}$

    Ⓑ $\frac{6}{12}$

    Ⓒ $\frac{5}{12}$

    Ⓓ $\frac{1}{4}$

2. **Find the sum:**

    $\frac{2}{7} + \frac{1}{7} =$

    Ⓐ $\frac{3}{7}$

    Ⓑ $\frac{3}{14}$

    Ⓒ $\frac{2}{14}$

    Ⓓ $\frac{1}{3}$

3. **What fraction expresses the total amount represented by the fraction models below?**

    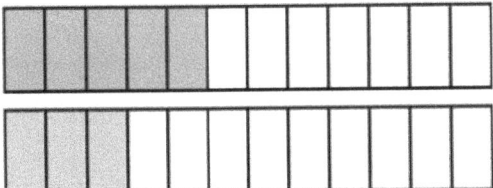

    Ⓐ $\frac{1}{2}$

    Ⓑ $\frac{5}{12}$

    Ⓒ $\frac{2}{3}$

    Ⓓ $\frac{3}{4}$

4. **What fraction expresses the total amount represented by the fraction models?**

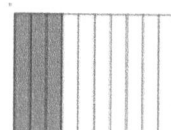 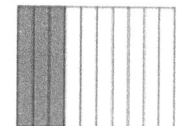

Ⓐ $\frac{3}{5}$

Ⓑ $\frac{3}{10}$

Ⓒ $\frac{1}{2}$

Ⓓ $\frac{3}{20}$

5. **Mary cut a pizza into 9 pieces. Mary ate 3 pieces, and Connie ate 3 pieces. What part of the pizza did they eat all together?**

Ⓐ $\frac{9}{18}$

Ⓑ $\frac{2}{3}$

Ⓒ $\frac{6}{18}$

Ⓓ $\frac{3}{4}$

6. **Write the simplified fraction that will complete the equation.**

$$\frac{1}{4} + \frac{5}{10} =$$

7. **Select the correct fractions to complete each equation.**

	$\frac{1}{2}$	$\frac{2}{5}$	$\frac{4}{9}$
$\frac{4}{5} - \frac{2}{5}$			
$\frac{1}{4} + \frac{1}{4}$			
$\frac{8}{9} - \frac{4}{9}$			

8. Write $2\frac{5}{8}$ as the sum of whole numbers and fractions. Select all the correct answers.

Ⓐ $2 + \frac{3}{8} + \frac{2}{8}$

Ⓑ $1 + \frac{4}{8} + \frac{1}{8}$

Ⓒ $1 + 1 + \frac{4}{8} + \frac{1}{8}$

Ⓓ $1 + \frac{2}{3} + \frac{3}{5}$

9. Find the difference. $7\frac{5}{9} - 5\frac{2}{9}$

Circle the correct answer.

Ⓐ $2\frac{4}{9}$

Ⓑ $2\frac{1}{9}$

Ⓒ $2\frac{5}{9}$

Ⓓ $2\frac{1}{3}$

CHAPTER 3 → Lesson 4: Adding and Subtracting Fractions through Decompositions

1. $1\frac{1}{4} - \frac{3}{4} = \square$

   (A) $\frac{6}{4} - \frac{3}{4} = \frac{3}{4}$

   (B) $\frac{4}{4} - \frac{3}{4} = \frac{1}{4}$

   (C) $\frac{4}{8} - \frac{3}{4} = \frac{1}{4}$

   (D) $\frac{5}{4} - \frac{3}{4} = \frac{2}{4} = \frac{1}{2}$

2. How many sevenths are there in 3 whole pizzas?

   (A) 7
   (B) 14
   (C) 28
   (D) 21

3. What fractional part could be added to each blank to make each number sentence true?

   $\frac{3}{8} = \frac{1}{8} + \underline{\quad} + \underline{\quad};$

   $\frac{3}{8} = \underline{\quad} + \frac{2}{8}$

   (A) $\frac{2}{8}$

   (B) $\frac{1}{8}$

   (C) $\frac{0}{8}$

   (D) $\frac{3}{8}$

4. $2\frac{3}{5} - \frac{4}{5} =$

   (A) $1\frac{1}{10}$

   (B) $1\frac{4}{5}$

   (C) $1\frac{2}{5}$

   (D) $1\frac{1}{5}$

5. How many sixths are there in 6 birthday cakes?

Ⓐ 30
Ⓑ 40
Ⓒ 36
Ⓓ 25

6. Match each fraction with its decomposition.

	$1\frac{2}{3}$	$2\frac{3}{5}$	$1\frac{1}{8}$
$1 + 1 + \frac{1}{5} + \frac{1}{5} + \frac{1}{5}$	◯	◯	◯
$1 + \frac{1}{3} + \frac{1}{3}$	◯	◯	◯
$1 + \frac{1}{8}$	◯	◯	◯

7. What does the following add up to? $1+1+1+\frac{1}{8}+\frac{1}{8}+\frac{1}{8}$
Write your answer in the box below.

8. Select all the correct answers for the following: $8\frac{3}{5} + \frac{4}{5}$

Ⓐ $8 + \frac{3}{5} + \frac{2}{5} + \frac{2}{5}$
Ⓑ $8 + \frac{7}{5}$
Ⓒ $8 + \frac{7}{10}$
Ⓓ $9 + \frac{2}{5}$

9. Solve $8\frac{1}{2} - 4\frac{3}{4}$.

Circle the correct answer.

Ⓐ $4\frac{3}{4}$
Ⓑ $3\frac{3}{4}$
Ⓒ $3\frac{1}{4}$
Ⓓ $4\frac{1}{4}$

CHAPTER 3 →Lesson 5: Adding and Subtracting Mixed Numbers

1. Angelo picked $2\frac{3}{4}$ pounds of apples from the apple orchard. He gave $1\frac{1}{4}$ pounds to his neighbor Mrs. Mason. How many pounds of apples does Angelo have left?

   Ⓐ $1\frac{1}{2}$ pounds

   Ⓑ $1\frac{3}{4}$ pounds

   Ⓒ $2\frac{1}{4}$ pounds

   Ⓓ $1\frac{3}{8}$ pounds

2. Daniel and Colby are building a castle out of plastic building blocks. They will need $2\frac{1}{2}$ buckets of blocks for the castle. Daniel used to have two full buckets of blocks, but lost some, and now only has $1\frac{3}{4}$ buckets. Colby used to have two full buckets of blocks too, but now has $1\frac{1}{4}$ buckets. If Daniel and Colby combine their buckets of blocks, will they have enough to build their castle?

   Ⓐ No, they will have less than $1\frac{1}{2}$ buckets.

   Ⓑ No, they will have $1\frac{1}{2}$ buckets.

   Ⓒ Yes, they will have $2\frac{1}{2}$ buckets.

   Ⓓ Yes, they will have 3 buckets.

3. Lexi and Ava are making chocolate chip cookies for a sleepover with their friends. They will need $4\frac{1}{4}$ cups of chocolate chips to make enough cookies for their friends. Lexi has $2\frac{3}{4}$ cups of chocolate chips. Ava has $1\frac{3}{4}$ cups of chocolate chips. Will the girls have enough chocolate chips to make the cookies for their friends?

   Ⓐ They'll have less than 4 cups, but should just use the amount they have.
   Ⓑ They'll have less than 4 cups, so no.
   Ⓒ They'll have $4\frac{1}{4}$ cups, so yes.
   Ⓓ They'll have $4\frac{2}{4}$ cups, so yes.

4. $3\frac{2}{4} + 1\frac{1}{4} =$

Ⓐ $\frac{20}{4}$

Ⓑ $2\frac{3}{4}$

Ⓒ $4\frac{3}{4}$

Ⓓ $5\frac{3}{4}$

5. $7\frac{9}{9} - 3\frac{5}{9} =$

Ⓐ $2\frac{7}{9}$

Ⓑ $3\frac{4}{9}$

Ⓒ $3\frac{3}{9}$

Ⓓ $4\frac{4}{9}$

6. What is $4\frac{2}{4} + 1\frac{2}{4}$?

7. Match each equation with the correct answer.

	$2\frac{2}{4}$	$3\frac{3}{4}$	$4\frac{1}{4}$
$2\frac{1}{4} + 1\frac{2}{4}$	○	○	○
$5\frac{1}{4} - 2\frac{3}{4}$	○	○	○
$2\frac{3}{4} + 1\frac{2}{4}$	○	○	○

8. $6\frac{3}{8} + 5\frac{7}{8}$ = ? Circle all the correct answers.

   Ⓐ $11\frac{10}{8}$

   Ⓑ $11\frac{10}{16}$

   Ⓒ $12\frac{1}{4}$

   Ⓓ $\frac{49}{4}$

9. John picked $2\frac{2}{3}$ pounds of apples. Together, John and Jose picked $4\frac{1}{2}$ pounds of apples. How many pounds of apples did Jose pick? Highlight the correct answer.

   Ⓐ $1\frac{5}{6}$

   Ⓑ $1\frac{1}{6}$

   Ⓒ $2\frac{5}{6}$

   Ⓓ $2\frac{1}{6}$

CHAPTER 3 → Lesson 6: Adding and Subtracting Fractions in Word Problems

1. Marcie and Lisa wanted to share a cheese pizza. Marcie ate $\frac{3}{6}$ of the pizza, and Lisa ate $\frac{2}{6}$ of the pizza. How much of the pizza did the girls eat together?

   (A) $\frac{6}{6}$ of a pizza

   (B) $\frac{5}{6}$ of a pizza

   (C) $\frac{1}{2}$ of a pizza

   (D) $\frac{4}{6}$ of a pizza

2. Sophie and Angie need $8\frac{5}{8}$ feet of ribbon to package gift baskets. Sophie has $3\frac{1}{8}$ feet of ribbon and Angie has $5\frac{3}{8}$ feet of ribbon. Will the girls have enough ribbon to complete the gift baskets?

   (A) Yes, and they will have extra ribbon.
   (B) Yes, but they will not have extra ribbon.
   (C) They will have just enough ribbon to make the baskets.
   (D) No, they will not have enough ribbon to make the baskets.

3. Travis has $4\frac{1}{8}$ pizzas left over from his soccer party. After giving some pizza to his friend, he has $2\frac{4}{8}$ of a pizza left. How much pizza did Travis give to his friend?

   (A) $1\frac{1}{2}$ pizzas

   (B) $1\frac{5}{8}$ pizzas

   (C) $1\frac{3}{4}$ pizzas

   (D) $1\frac{5}{7}$ pizzas

4. Which student solved the problem correctly?

Student 1	Student 2	Student 3
$3 + 2 = 5$ and $\frac{3}{4} + \frac{1}{4} = 1$ so $5 + 1 = 6$	$3\frac{3}{4} + 2 = 5\frac{3}{4}$ and $5\frac{3}{4} + \frac{1}{4} = 5\frac{4}{4} = 6$	$3\frac{3}{4} = \frac{15}{4}$ and $2\frac{1}{4} = \frac{9}{4}$ so $\frac{15}{4} + \frac{9}{4} = \frac{24}{4} = 6$

Ⓐ Student 1
Ⓑ Student 2
Ⓒ Student 3
Ⓓ All of the students

5. There are 2 loaves of freshly baked bread, and each loaf is cut into 8 equal pieces. If $\frac{5}{8}$ of a loaf is used for breakfast, and $\frac{7}{8}$ of a loaf is used for lunch, what fraction of the loaf if left?

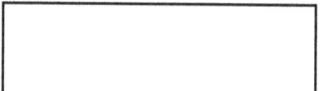 

Ⓐ $\frac{1}{2}$

Ⓑ $\frac{1}{16}$

Ⓒ $\frac{1}{8}$

Ⓓ $\frac{1}{10}$

6. Tom and Sally buy a box of chocolates. Tom eats $\frac{2}{7}$ of the chocolates. Sally eats $\frac{3}{7}$ of the chocolates. What fraction of chocolates did they eat altogether? Write your answer in the box given below.

7. Match each of the word problem with the correct answer

	$\frac{3}{5}$	$\frac{2}{3}$
There are two bags of candy. The first bag has $\frac{2}{5}$ of a bag, and the second has $\frac{1}{5}$ of a bag. What fraction of a bag of candy is there in all?	○	○
Izabel has a bag of marbles, but lets her brother have $\frac{1}{3}$ of them. What fraction represents how much she has left?	○	○

8. John and his friends ate $3\frac{1}{3}$ pizzas on Monday. They ate $2\frac{1}{4}$ pizzas on Tuesday. How much of the pizza did they eat in all? Select all the correct answers.

Ⓐ $5\frac{7}{12}$ pizzas

Ⓑ $5\frac{2}{7}$ pizzas

Ⓒ $\frac{37}{7}$ pizzas

Ⓓ $\frac{67}{12}$ pizzas

9. Jose had $5\frac{5}{12}$ m of ribbon. He used $3\frac{7}{8}$ m of the ribbon. How much ribbon did he have left? Mark the correct answer.

Ⓐ $2\frac{13}{24}$ m

Ⓑ $1\frac{3}{24}$ m

Ⓒ $1\frac{13}{24}$ m

Ⓓ $1\frac{9}{24}$ m

CHAPTER 3 →Lesson 7: Multiplying Fractions

1. Solve $\frac{1}{2}$ x 6 =

   Ⓐ $\frac{2}{6}$
   Ⓑ $\frac{1}{3}$
   Ⓒ 3
   Ⓓ $\frac{3}{6}$

2. Solve 6 x $\frac{1}{4}$ =

   Ⓐ 2
   Ⓑ $1\frac{1}{2}$
   Ⓒ $\frac{4}{6}$
   Ⓓ 3

3. What product do these models show?

   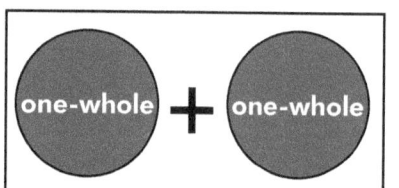 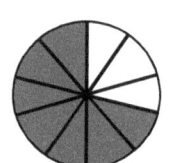

   Ⓐ $\frac{4}{10}$
   Ⓑ $\frac{14}{100}$
   Ⓒ $\frac{7}{10}$
   Ⓓ $1\frac{4}{10}$

4. Find the product:

   45 x $\frac{2}{3}$

   Ⓐ 90
   Ⓑ 86
   Ⓒ 30
   Ⓓ 129

5. Select all of the multiplication problems that will give us $\frac{1}{2}$ as our simplified answer.

Ⓐ $\frac{2}{3}$ x $\frac{2}{3}$

Ⓑ $\frac{2}{3}$ x $\frac{3}{4}$

Ⓒ $\frac{1}{2}$ x $\frac{10}{11}$

Ⓓ $\frac{3}{5}$ x $\frac{5}{6}$

Ⓔ $\frac{1}{4}$ x $\frac{2}{6}$

6. Write the fraction that will make the equation true. $\frac{1}{3}$ x $\frac{8}{10}$ Write your answer in the box given below.

7. Match each equation with its answer

	$\frac{8}{25}$	$\frac{1}{25}$	$\frac{1}{16}$
$\frac{4}{5} \times \frac{2}{5}$	○	○	○
$\frac{1}{4} \times \frac{1}{4}$	○	○	○
$\frac{2}{5} \times \frac{1}{10}$	○	○	○

8. Which of the following are equal to 5 x $\frac{2}{7}$. Select all the correct answers.

Ⓐ $\frac{10}{7}$

Ⓑ $\frac{2}{14}$

Ⓒ $1\frac{3}{7}$

Ⓓ $\frac{2}{7} + \frac{2}{7} + \frac{2}{7} + \frac{2}{7} + \frac{2}{7}$

CHAPTER 3 →Lesson 8: Multiplying Fractions by a Whole Number

1. $7 \times \frac{1}{3} =$

   Ⓐ $\frac{1}{21}$

   Ⓑ $\frac{3}{7}$

   Ⓒ $\frac{7}{22}$

   Ⓓ $\frac{7}{3}$

2. Cook is making sandwiches for the party. If each person at a party eats $\frac{2}{8}$ of a pound of turkey, and there are 5 people at the party, how many pounds of turkey are needed?

   Ⓐ 3 pounds

   Ⓑ $1\frac{1}{4}$ pounds

   Ⓒ $2\frac{1}{4}$ pounds

   Ⓓ $\frac{4}{8}$ pounds

3. $\frac{1}{5} \times 7 =$

   Ⓐ $\frac{7}{5}$

   Ⓑ $\frac{5}{7}$

   Ⓒ 35

   Ⓓ $\frac{1}{7}$

4. $\frac{1}{9} \times 8 =$

   Ⓐ 72

   Ⓑ $\frac{8}{9}$

   Ⓒ $\frac{1}{8}$

   Ⓓ $\frac{9}{8}$

5. Which product is greater than 1?

   Ⓐ $2 \times \frac{2}{3}$

   Ⓑ $4 \times \frac{1}{6}$

   Ⓒ $3 \times \frac{2}{10}$

   Ⓓ $2 \times \frac{2}{5}$

6. Fill in the missing number to complete the equation. $\frac{1}{2} \times$ ___ $= 4$

   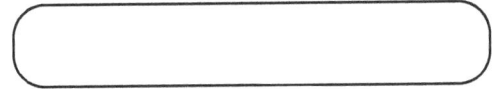

7. Match the problem with the correct answer

	$2\frac{2}{3}$	2	3
$(\frac{1}{3}) \times 6$	◯	◯	◯
$(\frac{2}{3}) \times 4$	◯	◯	◯
$(\frac{3}{4}) \times 4$	◯	◯	◯

8. Which of the following are equal to $3 \times \frac{7}{6}$. Select all the correct answers.

   Ⓐ $\frac{21}{6}$

   Ⓑ $\frac{10}{6}$

   Ⓒ $3\frac{1}{2}$

   Ⓓ $\frac{7}{6} + \frac{7}{6} + \frac{7}{6}$

CHAPTER 3 →Lesson 9: Multiplying Fractions in Word Problems

1. Aimee is making treat bags for her Christmas party. She is going to put $\frac{2}{3}$ cups of mint M&Ms in each bag. She has invited 9 friends to her party. How many cups of mint M&Ms does she need for her friends' treat bags?

   Ⓐ 5 cups
   Ⓑ 6 cups
   Ⓒ 4 cups
   Ⓓ 5 $\frac{1}{2}$ cups

2. Aimee was making treat bags from Question #1, but then she decided to also include $\frac{1}{2}$ of a cup of coconut M&Ms in each bag. How many cups of coconut M&Ms does she need? (Remember that she has 9 friends.)

   Ⓐ 4 $\frac{1}{2}$ cups
   Ⓑ 4 cups
   Ⓒ 3 $\frac{1}{3}$ cups
   Ⓓ 3 cups

3. Aimee's mom bought palm tree bags to celebrate their move to Florida. This is their first Christmas in Florida. Each treat bag will hold one cup of treats. Aimee wants to use $\frac{2}{3}$ cup mint M&M's and $\frac{1}{2}$ cup coconut M&M's. Will Aimee be able to fit all of the M&Ms in each party bag for her friends?

   Ⓐ Yes
   Ⓑ No

4. Kendra runs $\frac{3}{4}$ mile each day. How many miles does she run in 1 week?

   Ⓐ 5 miles
   Ⓑ 5 $\frac{1}{4}$ miles
   Ⓒ 5 $\frac{1}{2}$ miles
   Ⓓ 5 $\frac{3}{4}$ miles

5. Mrs. Howett is making a punch. The punch uses $\frac{3}{5}$ cup of grapefruit juice for one serving. If she makes 4 servings, how many cups of grapefruit juice does she need?

Ⓐ $2\frac{2}{5}$ cups

Ⓑ $1\frac{2}{5}$ cups

Ⓒ $3\frac{1}{5}$ cups

Ⓓ $1\frac{4}{5}$ cups

6. Tim has 3 cups of milk. He used 1/3 of the milk. How many cups of milk are left?

7. Match each of the problem with the correct answer choice.

	$40	$60
The dinner for a large family costs $80. Mr. Smith has a $\frac{1}{2}$ off coupon, so what will the final price be?	○	○
The movie tickets cost $90 but I have a coupon for $\frac{1}{3}$ off. How much will I spend?	○	○

8. John can paint $\frac{2}{5}$ of a table in 15 minutes. Jose can paint 8 times that amount in 15 minutes. How many tables can Jose paint in 15 minutes? Circle the correct answer.

Ⓐ $2\frac{3}{5}$

Ⓑ $\frac{1}{20}$

Ⓒ $3\frac{2}{5}$

Ⓓ $3\frac{1}{5}$

CHAPTER 3 →Lesson 10: 10 to 100 Equivalent Fractions

1. **What fraction of a dollar is 6 dimes and 3 pennies?**
   Ⓐ 0.63 or $\frac{63}{100}$

   Ⓑ 0.73 or $\frac{73}{100}$

   Ⓒ 0.53 or $\frac{53}{100}$

   Ⓓ 0.43 or $\frac{43}{100}$

2. **1 tenth + 4 hundredths = _____ hundredths**

   Ⓐ 14
   Ⓑ 140
   Ⓒ 1400
   Ⓓ 104

3. **5 tenths + 2 hundredths = _____ hundredths**

   Ⓐ 25
   Ⓑ 525
   Ⓒ 52
   Ⓓ 502

4. **5 hundredths + 2 tenths = _____ hundredths**

   Ⓐ 25
   Ⓑ 52
   Ⓒ 252
   Ⓓ 502

5. **14 hundredths = _____ hundredths + 4 hundredths**

   Ⓐ 144
   Ⓑ 414
   Ⓒ 104
   Ⓓ 10

6. 14 hundredths = _____ tenth(s) + 4 hundredths

   (A) 10
   (B) 100
   (C) 1
   (D) 0

7. 14 hundredths = 1 tenth + 3 hundredths + _____ hundredth(s)

   (A) 10
   (B) 1
   (C) 0
   (D) 4

8. 80 hundredths = _____ tenths

   (A) 8
   (B) 80
   (C) 0
   (D) 1

9. $\frac{2}{10} + \frac{41}{100} =$

   (A) $\frac{43}{100}$

   (B) $\frac{43}{10}$

   (C) $\frac{61}{100}$

   (D) $\frac{61}{10}$

10. Match each fraction to its equivalent.

	$\frac{10}{100}$	$\frac{60}{100}$	$\frac{90}{100}$
$\frac{9}{10}$	○	○	○
$\frac{1}{10}$	○	○	○
$\frac{6}{10}$	○	○	○

11. Select the equivalent fraction of $\frac{2}{10}$ that has a denominator of 100. Circle the correct answer.

 Ⓐ $\frac{12}{100}$

 Ⓑ $\frac{12}{100}$

 Ⓒ $\frac{20}{100}$

 Ⓓ $\frac{1}{5}$

12. Write the equivalent fraction of $\frac{4}{10}$ that now has a denominator of 100. Instruction: Write in the format $\frac{A}{B}$.

13. Solve $8\frac{5}{100} - 5$

 Ⓐ $2\frac{75}{100}$

 Ⓑ $3\frac{75}{100}$

 Ⓒ $2\frac{3}{4}$

 Ⓓ $3\frac{3}{4}$

CHAPTER 3 →Lesson 11: Convert Fractions to Decimals

1. **Point A is located closest to _____ on this number line.**

    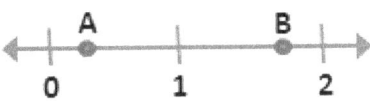

    Ⓐ 0
    Ⓑ 0.75
    Ⓒ 0.25
    Ⓓ -1

2. **Convert $\frac{148}{1000}$ to a decimal.**

    Ⓐ 0.148
    Ⓑ 14.8
    Ⓒ .00148
    Ⓓ 0.0148

3. **Where is Point D located on this number line?**

    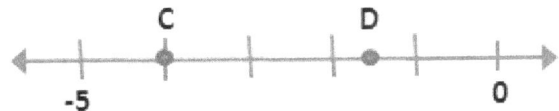

    Ⓐ -2.5
    Ⓑ -2
    Ⓒ -1.5
    Ⓓ -3

4. **Convert $129\frac{1}{4}$ to decimal.**

    Ⓐ 129.25
    Ⓑ 129.025
    Ⓒ 129.75
    Ⓓ 129.14

5. **Convert the fraction to a decimal: $\frac{3}{1000}$ = _____**

    Ⓐ 0.03
    Ⓑ 0.003
    Ⓒ 0.30
    Ⓓ 0.0003

6. What decimal does this model represent?

Ⓐ 2.7
Ⓑ 2.0
Ⓒ 0.207
Ⓓ 0.27

7. Mr. Hughes preferred to convert this decimal into a fraction in order to add it to a group of other fractions. Which answer is correct?

0.044

Ⓐ $\frac{44}{100}$

Ⓑ $\frac{44}{10000}$

Ⓒ $\frac{44}{1000}$

Ⓓ $\frac{44}{10}$

8. Convert the mixed number to a decimal: $1\frac{4}{1000}$ = _____

Ⓐ 1.04
Ⓑ 1.004
Ⓒ 1.4
Ⓓ .1004

9. The construction crew was working 0.193 of the times that we traveled that road. What fraction of the time was the crew working?

Ⓐ $\frac{193}{100}$

Ⓑ $\frac{193}{10000}$

Ⓒ $\frac{193}{1000}$

Ⓓ $\frac{193}{10}$

10. What are the addends in this problem?
0.300 + 0.249

Ⓐ $\frac{300}{100} + \frac{249}{100}$

Ⓑ $\frac{300}{100} + \frac{249}{1000}$

Ⓒ $\frac{549}{1000}$

Ⓓ $\frac{300}{1000} + \frac{249}{1000}$

11. Match each fraction to its decimal.

	0.50	0.25	0.10
$\frac{1}{2}$	○	○	○
$\frac{1}{4}$	○	○	○
$\frac{1}{10}$	○	○	○

12. Circle all the fractions that are equivalent to 0.25.

Ⓐ $\frac{1}{2}$

Ⓑ $\frac{1}{4}$

Ⓒ $\frac{9}{10}$

Ⓓ $\frac{3}{5}$

Ⓔ $\frac{25}{100}$

13. Write the decimal that is equivalent to $\frac{1}{20}$. Write your answer in the box below.

14. Which of the following numbers are equal to $\frac{6}{10}$? Select all the correct answers.

Ⓐ 0.6
Ⓑ 0.06
Ⓒ $\frac{60}{100}$
Ⓓ 0.60

CHAPTER 3 →Lesson 12: Compare Decimals

1. **Point B is located closest to ____ on this number line.**

   Ⓐ 1.75
   Ⓑ 1.5
   Ⓒ 2.25
   Ⓓ 1.1

2. **Compare the following decimals using <, >, or =.**
   0.05 ___ 0.50

   Ⓐ 0.05 < 0.50
   Ⓑ 0.05 = 0.50
   Ⓒ 0.05 > 0.50

3. **The rainbow trout dinner costs $23.99. The steak dinner costs $26.49. The roast chicken dinner is $5.00 less than $30.79. Which dinner costs the most?**

   Ⓐ rainbow trout dinner
   Ⓑ steak dinner
   Ⓒ chicken dinner

4. **Compare the following decimals using <, >, or =.**
   0.2 ____ 0.200

   Ⓐ 0.2 < 0.200
   Ⓑ 0.2 = 0.200
   Ⓒ 0.2 > 0.200

5. **Which of these decimals is the greatest?**

   Ⓐ 0.0060
   Ⓑ 0.006
   Ⓒ 0.060

6. **Which comparison symbol makes this statement true?**
   1.954 ___ 0.1954

   Ⓐ 1.954 = 0.1954
   Ⓑ 1.954 > 0.1954
   Ⓒ 1.954 < 0.1954

7. **Order these decimals from least to greatest.**
   43.75; 0.4385; 0.04375

   Ⓐ 0.04375; 43.75; 0.4385
   Ⓑ 0.4385; 0.04375; 43.75
   Ⓒ 0.04375; 0.4385; 43.75
   Ⓓ 0.4385; 43.75; 0.04375

8. **Compare the following decimals using <, >, or =.**
   1.10 ___ 1.1000

   Ⓐ 1.10 < 1.1000
   Ⓑ 1.10 = 1.1000
   Ⓒ 1.10 > 1.1000

9. **A one-way airline ticket to Atlanta from New York costs $65.00 more on the weekend than it does during the week. How much would a $225 (weekday price) ticket cost if the traveler needed to fly on a Saturday?**

   Ⓐ $270.00
   Ⓑ $290.00
   Ⓒ $280.00
   Ⓓ $300.00

10. A pattern exists in the prices of the following vehicles. Which numbers complete this table?

Vehicle	Price
compact car	$30,000.00
midsize car	$35,000.00
luxury car	$40,000.00
jeep	?
mini van	$50,000.00
full size van	?

Ⓐ $36,000.00; $51,000.00
Ⓑ $45,000.00; $55,000.00
Ⓒ $40,000.00; $51,000.00
Ⓓ $36,000.00; $55,000.00

11. Observe the decimals numbers given below.
0.93, 0.39, 0.84, 0.09
Which number is the largest ? Identify the number and write it in the box given below

12. Match the number pair with the correct inequality sign.

	<	>	=
0.25 _ 0.39		○	○
0.89 _ 0.890	○	○	○
0.12 _ 0.21	○	○	○
0.29 _ 0.28	○	○	○

13. Circle all the decimals that are less than 0.45.

0.28, 0.96, 0.39, 0.42 , 0.58, 0.04, 0.37, 0.49

14. Two decimal numbers X and Y are represented by the two area models below. Compare the two numbers. From among the 4 options given below, identify the correct statement and write it in the box given below.

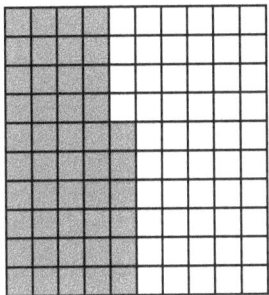 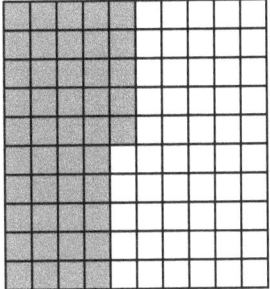

Ⓐ x > y
Ⓑ x = y
Ⓒ x < y

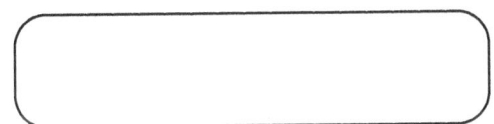

15. Compare the first number with the second number in each row and fill the table by writing the correct inequality sign in the table.

First Number	Symbol	Second Number
8 ones and 5 hundredths	<	8.5
$7\frac{56}{10}$		7.56
5 tenths		25 hundredths
6.52		65 tenths and 2 hundredths

End of Number & Operations - Fractions

Chapter 3
Number & Operations - Fractions
Lesson 1: Equivalent Fractions

Question No.	Answer	Detailed Explanation
1	C	The denominator (bottom number) is the total number of items presented. The numerator (top number) is the number of identified items.
2	C	There are three different shapes represented. This question is asking for the number of squares and triangles. That number of shapes that are not circles is the numerator and the total number of shapes is the denominator.
3	A	The fraction should only pertain to the number of squares: the number of shaded squares is the numerator and the total number of squares is the denominator.
4	D	The number of shaded circles is the numerator and the total number of shaded shapes is the denominator.
5	C	All of these fractions represent $\frac{1}{2}$. The numerators are 1 part out of 2 parts: 4 is two parts of 2. 6 is two parts of 3.
6	B	The correct answer would be fractions which have numerators with a sum of 11 and denominators that are both 12.
7	D	The model represents 1 part of something that is divided into 4 equal pieces. An equivalent fraction would also be $\frac{1}{4}$ of a total number of parts. $\frac{4}{16} = \frac{1}{4}$. Hence, the correct answer is D.
8	D	Draw a model of $\frac{8}{18}$. Choose the fraction that has the same portion sizes as $\frac{8}{18}$.
9	B	Each equivalent fraction represents 5 parts out of 6. When we multiply both numerators and denominators by a common factor, the new fraction will be equivalent to $\frac{5}{6}$. For eg. $\frac{5 \times 2}{6 \times 2} = \frac{10}{12}$ is equivalent to $\frac{5}{6}$. So, among the options, we see that option (B) $= \frac{20}{24}$ is correct. Because $\frac{20}{24}$ reduces to $\frac{5}{6}$, when the common factor is canceled. $\frac{20}{24} = \frac{5 \times 4}{6 \times 4} = \frac{5}{6}$.

Question No.	Answer	Detailed Explanation
9	B	Each equivalent fraction represents 5 parts out of 6. When we multiply both numerators and denominators by a common factor, the new fraction will be equivalent to $\frac{5}{6}$. For eg. $\frac{5 \times 2}{6 \times 2} = \frac{10}{12}$ is equivalent to $\frac{5}{6}$. So, among the options, we see that option (B) $= \frac{20}{24}$ is correct. Because $\frac{20}{24}$ reduces to $\frac{5}{6}$, when the common factor is canceled. $\frac{20}{24} = \frac{5 \times 4}{6 \times 4} = \frac{5}{6}$.
10	B	Find the GCF. This is the largest number that both the numerator and denominator can be divided by. The quotients are the numerator and denominator reduced to its lowest terms: for example, $\frac{15}{20}$ is reduced to $\frac{3}{4}$ because 15 is divided by 5 (GCF) 3 times and 20, 4 times. Five is the largest number that 15 and 20 can be divided by evenly. In our problem, $\frac{21}{49}$ can be reduced to $\frac{3}{7}$, because 21 is divided by 7 (GCF) 3 times and 49, 7 times.
11	D	Find the GCF, which is the largest factor that both the numerator and denominator can be divided by.
12	B	The correct fraction can be reduced to its lowest terms of $\frac{4}{5}$: Find the Greatest Common Factor (GCF), which is a number that the numerator and denominator can be divided by: 80 divided by 20 = 4 and 100 divided by 20 = 5. In this case, the GCF is 20. The number of times the numerator and denominator divides evenly into the GCF ($\frac{4}{5}$) is the lowest terms. $\frac{60}{75}$ also reduces to $\frac{4}{5}$ when reduced to lowest terms. (GCF = 15)
13	B	Reduce the fraction to its lowest terms by dividing the numerator and denominator by the GCF (9).
14	C	Use the GCF of the numerator and denominator of each fraction to determine if it is equivalent to $\frac{2}{9}$.
15	B	These fractions all reduce to $\frac{5}{8}$ in their lowest terms.

16

	Equivalent	Not Equivalent
$\frac{12}{15}$ and $\frac{3}{5}$		○
$\frac{18}{24}$ and $\frac{9}{12}$	○	
$\frac{18}{200}$ and $\frac{9}{100}$	○	
$\frac{3}{15}$ and $\frac{3}{25}$		○

To find if the fractions are equivalent, change both of them into their simplest form. If the simplest form is the same, they are equivalent fractions. For example, 18/24 and 9/12 can be reduced to 3/4. So 18/24 and 9/12 are equivalent fractions. If the simplest forms are not the same, then the fractions are not equivalent. For example, 12/15 reduces to 4/5. So, 12/15 and 3/5 are not equivalent.

17   4/5

30 is the GCF of 120 and 150. When the GCF is taken out from both the numerator and denominator, 120/150 reduces to 4/5.

18   B,D,F

Divide out common terms as much as you can. Once you cannot simplify anymore, see which fractions are equivalent to $\frac{1}{2}$.

$$\frac{2}{4} = \frac{^2/_2}{^4/_2} = \frac{1}{2}$$

$$\frac{35}{70} = \frac{^{35}/_{35}}{^{70}/_{35}} = \frac{1}{2}$$

$$\frac{7}{14} = \frac{^7/_7}{^{14}/_7} = \frac{1}{2}$$

Therefore, $\frac{2}{4}$, $\frac{35}{70}$ and $\frac{7}{14}$ are equivalent to $\frac{1}{2}$

19   B & D

$\frac{1}{3} = \frac{1 \times 2}{3 \times 2} = \frac{2}{6}$; $\frac{1}{3} = \frac{1 \times 3}{3 \times 3} = \frac{3}{9}$; Therefore, option (B) is correct.

$\frac{1}{3} = \frac{1 \times 6}{3 \times 6} = \frac{6}{18}$; $\frac{1}{3} = \frac{1 \times 12}{3 \times 12} = \frac{12}{36}$; $\frac{1}{3} = \frac{1 \times 15}{3 \times 15} = \frac{15}{45}$

Therefore, option (D) is correct.

Lesson 2: Compare Fractions

Question No.	Answer	Detailed Explanation
1	C	Point D is located halfway between 3 and 4 on the number line. This point would represent 3.5 or $3\frac{1}{2}$.
2	A	All of these fractions are equivalent because they reduce to the same fraction when reduced to lowest terms. They would all reduce to $\frac{2}{7}$.
3	C	Both fractions are showing three parts. $\frac{3}{12}$ is 3 parts out of 12; $\frac{3}{18}$ is 3 parts out of 18. The greater the denominator, the more pieces the wholes have been divided into. Therefore, the pieces will be smaller.
4	B	Four pieces from a cake that has been divided into 20 equal pieces will be more cake than 4 pieces cut from a cake divided into 28 equal pieces.
5	A	Add the numerators to compare the actual sum with the third fraction.
6	A	Subtract the numerators to compare the numerator with the third fraction.
7	C	The model above is divided into 12 equal parts. If written as a fraction, the denominator would be 12. The greater the denominator, the more pieces the pie or whole is divided into, and therefore, the smaller the pieces. $\frac{3}{10}$ would be greater than $\frac{2}{12}$.
8	B	The fractions should get smaller. To compare the fractions, first convert them into equivalent fractions with the same denominator.
9	B	The shaded area in Model C is the greatest. The shaded area in Model B is the least.
10	A	Add the numerators and compare the sums.
11	A	The denominators of these fractions are all the same, so the numerators should be listed from least to greatest. 13 could go between 9 and 16.
12	C	Subtract $\frac{3}{7}$ from $\frac{6}{7}$. $\frac{6}{7} - \frac{3}{7} = \frac{3}{7}$. Dan eats vegetables $\frac{3}{7}$ of a week (or 3 days) less than Marty.
13	C	$\frac{1}{2}$ is the only fraction listed that is greater than $\frac{1}{3}$.
14	B	The denominators are all the same, so use the numerators of the fractions to list them in increasing order.
15	C	If a whole is divided into 9 equal pieces and another congruent whole is divided into 11 equal pieces, the elevenths will be smaller than the ninths.

Question No.	Answer	Detailed Explanation

16

		<	>	=
$\frac{24}{36}$	$\frac{2}{4}$		◯	
$\frac{1}{4}$	$\frac{5}{11}$	◯		
$\frac{10}{20}$	$\frac{7}{14}$			◯
$\frac{4}{4}$	$\frac{10}{11}$		◯	

Looking at the fractions, divide out common terms as much as you can. Once you cannot simplify anymore, see which fraction is greater. That will get the two points in the inequality. If each fraction is the same when simplified, the fractions are equal.

The first row is >, the second row is <, the third row is =, and the last row is >.

17 — **A,B,F**

Looking at the fractions, divide out common terms as much as you can. Once you cannot simplify anymore, see which fractions are greater than 2/3. The correct answers you should have clicked on are 30/36, 3/4, and 7/9.

18. — **B & C**

To compare two fractions, we write equivalent fractions of them which have the same numerator or denominator.

$\frac{3}{5} < \frac{4}{5}$ (When a whole is divided into 5 equal parts, 3 parts are less than 4 parts). Therefore, option (A) is wrong.

$\frac{1}{5} > \frac{1}{10}$. Because $\frac{1}{5}$ is one part when the whole is divided into 5 equal parts and $\frac{1}{10}$ is one part when the congruent whole is divided into 10 equal parts. Since $5 < 10$, $\frac{1}{5} > \frac{1}{10}$. So, 3 fifths ($\frac{3}{5}$) is more than 3 tenths ($\frac{3}{10}$). $\frac{3}{5} > \frac{3}{10}$. Therefore, option (B) is correct.

$\frac{3}{5} = \frac{3 \times 3}{5 \times 3} = \frac{9}{15}$. $\frac{9}{15} > \frac{8}{15}$ (When a whole is divided into 15 equal parts, 9 parts are more than 8 parts). or $\frac{3}{5} > \frac{8}{15}$. Therefore, option (C) is correct.
$\frac{3}{5} = \frac{(3 \times 4)}{(5 \times 4)} = \frac{12}{20}$. Therefore, option (D) is incorrect.

Lesson 3: Adding and Subtracting Fractions

Question No.	Answer	Detailed Explanation
1	C	When denominators are the same, showing a portion that is divided into the same number of parts, add the numerators. The numerator is the sum of the numerators, and the denominator remains the same.
2	A	The denominators are the same, so add the numerators.
3	C	The shaded sections represent the numerator. The number of sections the models are divided up into (counted one time) is the denominator. In the first row, there are $\frac{5}{12}$ shaded, and in the second row, there are $\frac{3}{12}$ shaded which equals $\frac{8}{12}$. 8 can be divided by 4 which equals 2. 12 can be divided by 4 which equals 3. Therefore, $\frac{8}{12}$ can be reduced to $\frac{2}{3}$. Reduce the fraction to its lowest terms.
4	A	Add up all of the shaded areas; this is the numerator. The denominator is the number of sections a model is divided into. $\frac{3}{10}$ plus $\frac{3}{10}$ would need to be added together. Since the denominator is the same, only the numerator needs to be added (6). So, the answer is $\frac{6}{10}$; GCF of 6 and 10 is 2. When both the numerator and denominator are divided by 2, we get $\frac{3}{5}$.
5	B	The number of pieces the pizza is cut into is 9, which is the denominator. Each girl ate 3 pieces. The total number of pieces eaten is the numerator, which is 6. $\frac{6}{9}$ is the answer. $\frac{6}{9}$ can be divided evenly by 3, so the answer is $\frac{2}{3}$. Reduce the fraction to its lowest terms.
6	$\frac{3}{4}$	To add fractions, you have to start with a common denominator. Looking at 4 and 10, the LCM (Lowest Common Multiple) is 20. Write both the fractions with 20 as the denominator. $$\frac{1}{4} = \frac{(1\times5)}{(4\times5)} = \frac{5}{20}$$ $$\frac{5}{10} = \frac{(5\times2)}{(10\times2)} = \frac{10}{20}.$$ $\frac{5}{20} + \frac{10}{20} = \frac{15}{20}$. 5 is the GCF of 15 and 20. By taking out GCF (5) from both the numerator and denominator, $\frac{15}{20}$ reduces to $\frac{3}{4}$.

7

	$\frac{1}{2}$	$\frac{2}{5}$	$\frac{4}{9}$
$\frac{4}{5}$ - $\frac{2}{5}$		●	
$\frac{1}{4}$ + $\frac{1}{4}$	●		
$\frac{8}{9}$ - $\frac{4}{9}$			●

To add fractions, you have to start with a common denominator. In this case, we have common denominators, so we just add/subtract the numerators and leave the denominators the same.

$$\frac{4}{5} - \frac{2}{5} = \frac{2}{5}$$

$$\frac{1}{4} + \frac{1}{4} = \frac{1}{2}$$

$$\frac{8}{9} - \frac{4}{9} = \frac{4}{9}$$

8   A & C

$2 + \frac{3}{8} + \frac{2}{8} = 2 + \frac{3+2}{8} = 2\frac{5}{8}$. Therefore, option A is correct.

$1 + \frac{4}{8} + \frac{1}{8} = 1 + \frac{4+1}{8} = 1\frac{5}{8}$. Therefore, option B is wrong.

$1 + 1 + \frac{4}{8} + \frac{1}{8} = 2 + \frac{4+1}{8} = 2\frac{5}{8}$. Therefore, option C is correct.

$\frac{2}{3} + \frac{3}{5} = \frac{2 \times 5}{3 \times 5} + \frac{3 \times 3}{5 \times 3} = \frac{10}{15} + \frac{9}{15} = \frac{10+9}{15} = \frac{19}{15}$ Option D$= 2\frac{19}{15}$.

Therefore, option D is wrong.

9   D

Subtract the whole numbers and fractions separately and then simplify if possible.

$$7\frac{5}{9} - 5\frac{2}{9} = (7 + \frac{5}{9}) - (5 + \frac{2}{9})$$

$$= (7-5) + (\frac{5}{9} - \frac{2}{9}) = 2 + \frac{5-2}{9} = 2 + \frac{3}{9} = 2 + \frac{1}{3} = 2\frac{1}{3}$$

Lesson 4: Adding and Subtracting Fractions through Decompositions

Question No.	Answer	Detailed Explanation
1	D	One whole would be $\frac{4}{4}$, and you then add an additional $\frac{1}{4}$, for a total of $\frac{5}{4}$. Then, subtract $\frac{3}{4}$ from $\frac{5}{4}$. Reduce it to $\frac{1}{2}$.
2	D	One whole would be $\frac{7}{7}$. Two wholes would be $\frac{7}{7} + \frac{7}{7} = \frac{14}{7}$. Three whole would be $\frac{21}{7}$.
3	B	Each number sentence should equal $\frac{3}{8}$. In the first sentence, $\frac{1}{8}$ needs to be added twice, and in the second sentence once.
4	B	Explanation: $2\frac{3}{5}$ is equal to $\frac{13}{5}$. $\frac{13}{5} - \frac{4}{5}$ equals $\frac{9}{5}$, or, $1\frac{4}{5}$.
5	C	6 cakes multiplied by 6 pieces in each cake would be 36 pieces

6

	$1\frac{2}{3}$	$2\frac{3}{5}$	$1\frac{1}{8}$
$1 + 1 + \frac{1}{5} + \frac{1}{5} + \frac{1}{5}$		◯	
$1 + \frac{1}{3} + \frac{1}{3}$	◯		
$1 + \frac{1}{8}$			◯

Add the whole number and then add the fractions. To add fractions, you first have to have the denominator the same. After that, add the numerators and keep the denominator the same. Finally, reduce if possible.

7.	$3\frac{3}{8}$	Add the whole number and then add the fractions. To add fractions, you first have to have the denominator the same. After that, add the numerators and keep the denominator the same. Finally, reduce if possible.

Question No.	Answer	Detailed Explanation
8	A,B,D	$\frac{4}{5}$ can be decomposed as $\frac{2}{5} + \frac{2}{5}$. Therefore, option (A) is correct. $8 + \frac{3}{5} + \frac{2}{5} + \frac{2}{5} = 8 + \frac{3+2+2}{5} = \frac{7}{5}$. Therefore, option (B) is correct. $8 + \frac{7}{5} = 8 + \frac{5+2}{5} = 8 + \frac{5}{5} + \frac{2}{5} = 8 + 1 + \frac{2}{5} = 9 + \frac{2}{5}$. Therefore, option (D) is correct.
9	B	Since $\frac{1}{2} < \frac{3}{4}$, we regroup the first mixed number. $8\frac{1}{2} = 7 + \frac{4}{4} + \frac{1}{2} = 7 + \frac{4}{4} + \frac{1 \times 2}{2 \times 2} = 7 + \frac{4}{4} + \frac{2}{4}$ $= 7 + \frac{4+2}{4} = 7\frac{6}{4}$ Next, we subtract the whole numbers and fractions separately and then simplify if possible. $8\frac{1}{2} - 4\frac{3}{4} = 7\frac{6}{4} - 4\frac{3}{4} = \left(7 + \frac{6}{4}\right) - \left(4 + \frac{3}{4}\right)$ $= (7-4) + \left(\frac{6}{4} - \frac{3}{4}\right) = 3 + \frac{6-3}{4} = 3 + \frac{3}{4} = 3\frac{3}{4}$

Lesson 5: Adding and Subtracting Mixed Numbers

Question No.	Answer	Detailed Explanation
1	A	To determine how many pounds of apples he has left, subtract the amount of apples he gave away from the amount of apples he picked. Break this into two smaller subtraction problems, beginning with the whole numbers. Angelo had 2 full pounds of apples, and gave 1 full pound away: $2-1=1$. He now has 1 full pound of apples. In addition to the 2 full pounds, he had 3/4 of a pound of apples, and gave $\frac{1}{4}$ of a pound away: $\frac{3}{4} - \frac{1}{4} = \frac{2}{4}$. He now has $\frac{2}{4}$ of a pound of apples in addition to the 1 full pound of apples. All together, Angelo has $1\frac{2}{4}$ pounds of apples left. This can be reduced to $1\frac{1}{2}$ pounds.
2	D	$1\frac{3}{4} + 1\frac{1}{4} = 1 + \frac{3}{4} + 1 + \frac{1}{4}$. Using the Commutative Property of Addition, we have $1 + \frac{3}{4} + 1 + \frac{1}{4} = 1 + 1 + \frac{3}{4} + \frac{1}{4}$. We can then simply add the whole numbers together and add the fractions together. Adding the whole numbers we have $1 + 1 = 2$. Adding the fractions we have $\frac{3}{4} + \frac{1}{4} = \frac{4}{4} = 1$. Together we have $2+1=3$. Combined, the boys have 3 buckets of plastic building blocks.
3	D	The amount of chocolate chips needed can be expressed in number of fractional pieces ($\frac{17}{4}$), or number of whole, and additional fractional pieces ($4\frac{1}{4}$). This task requires an understanding of comparing fractions or mixed numbers. Combine $2\frac{3}{4}$ and $1\frac{3}{4}$. $2 + 1 = 3$ and $\frac{3}{4} + \frac{3}{4} = \frac{6}{4} = 1\frac{2}{4}$. The combined total of chips will be $4\frac{2}{4}$ cups. Because $4\frac{2}{4}$ is greater than $4\frac{1}{4}$, Lexi and Ava will have enough chocolate chips for their cookies. $4\frac{2}{4}$ can be reduced to $4\frac{1}{2}$.
4	C	Add the whole numbers (3) and (1) to get (4). Then add $\frac{2}{4}$ and $\frac{1}{4}$ to equal $\frac{3}{4}$.
5	D	Subtract whole numbers (7) – (3) for an answer of 4. To subtract fractions: Since the denominators are same, subtract numerators to get the numerator of the result. and denominator remains the same. Fractional part = $\frac{9-5}{9} = \frac{4}{9}$. Combine answers to get the final result: $4\frac{4}{9}$

Question No.	Answer	Detailed Explanation
6	6	First, add the whole numbers together, and then add the fractions. To add fractions, you first have to have the denominator the same. After that, add the numerators and keep the denominator the same. Finally, reduce if possible

7			$2\frac{2}{4}$	$3\frac{3}{4}$	$4\frac{1}{4}$
		$2\frac{1}{4} + 1\frac{2}{4}$		●	
		$5\frac{1}{4} - 2\frac{3}{4}$	●		
		$2\frac{3}{4} + 1\frac{2}{4}$			●

First, add the whole numbers together, and then add the fractions. To add fractions, you first have to have the denominator the same. After that, add the numerators and keep the denominator the same. Finally, reduce if possible. Subtraction: First, look at the fractions, and see if you have to regroup. Then subtract the fractions, by making sure you have the same denominator, subtracting the numerators, and keeping the denominator the same. Finally, subtract the whole numbers.

| 8 | A,C & D | $6\frac{3}{8} + 5\frac{7}{8} = (6+5) + \left(\frac{3}{8} + \frac{7}{8}\right) = 11 + \frac{3+7}{8} = 11\frac{10}{8}$ |

Therefore, option (A) is correct.

$$11\frac{10}{8} = 11 + \frac{10}{8} = 11 + \frac{8+2}{8} = 11 + \frac{8}{8} + \frac{2}{8} = 11 + 1 + \frac{2}{8}$$
$$= 12 + \frac{2 \div 2}{8 \div 2} = 12 + \frac{1}{4} = 12\frac{1}{4}$$

To convert a mixed number to an improper fraction, multiply the denominator by the whole number, then add the numerator. This sum becomes the numerator of the improper fraction. The denominator of the improper fraction is the same as the denominator of the fraction part of the mixed number.

$$12\frac{1}{4} = \frac{(4 \times 12) + 1}{4} = \frac{48+1}{4} = \frac{49}{4}$$

Therefore, option (D) is correct.

Question No.	Answer	Detailed Explanation
9	A	We subtract the weight of apples John picked from the total weight of apples to get the weight apples picked by Jose.

Subtract the whole numbers and fractions separately and then put them together.

Since the denominators of the fraction parts are different, write the equivalent fractions of them with the least common denominator(LCD).

LCD of 2 and 3 is 6.

$$\frac{1}{2} = \frac{1 \times 3}{2 \times 3} = \frac{3}{6}; \quad \frac{2}{2} = \frac{2 \times 2}{3 \times 2} = \frac{4}{6}$$

Since $\frac{3}{6} < \frac{4}{6}$, we regroup the first mixed number.

$$4\frac{1}{2} = 4\frac{3}{6} = 3 + \frac{6}{6} + \frac{3}{6} = 3 + \frac{6+3}{6} = 3 + \frac{9}{6}$$

$$4\frac{1}{2} - 2\frac{2}{3} = 4\frac{3}{6} - 2\frac{4}{6} = 3 + \frac{9}{6} - (2 + \frac{4}{6})$$

$$= (3-2) + (\frac{9}{6} - \frac{4}{6}) = 1 + \frac{9-4}{6} = 1 + \frac{5}{6} = 1\frac{5}{6}$$

Lesson 6: Adding and Subtracting Fractions in Word Problems

Question No.	Answer	Detailed Explanation
1	B	The amount of pizza Marcie ate can be thought of as $\frac{3}{6}$ (or $\frac{1}{6}$ and $\frac{1}{6}$ and $\frac{1}{6}$). The amount of pizza Lisa ate can be thought of as $\frac{1}{6}$ and $\frac{1}{6}$. The total amount of pizza they ate is $\frac{1}{6} + \frac{1}{6} + \frac{1}{6} + \frac{1}{6} + \frac{1}{6}$ or $\frac{5}{6}$ of the whole pizza.
2	D	The ribbon Sophie has can be added to the ribbon Angie has to determine how much ribbon they have altogether. Sophie has $3\frac{1}{8}$ feet of ribbon and Angie has $5\frac{3}{8}$ feet of ribbon. This can be written as $3\frac{1}{8} + 5\frac{3}{8}$. You add the 3 and 5, and they also have $\frac{1}{8}$ and $\frac{3}{8}$ which makes a total of $\frac{4}{8}$ more. Altogether the girls have $8\frac{4}{8}$ feet of ribbon, $8\frac{4}{8}$ is less than $8\frac{5}{8}$ so they will not have enough ribbon to complete the project. They will be short by $\frac{1}{8}$ of a foot of ribbon.
3	B	Travis had $4\frac{1}{8}$ pizzas to start. This is $\frac{33}{8}$ of a pizza. The x's show the pizza he has left which is $2\frac{4}{8}$ pizzas or $\frac{20}{8}$ pizzas. The shaded rectangles without the x's are the pizza he gave to his friend which is $\frac{13}{8}$ or $1\frac{5}{8}$ pizzas.
4	D	All ways are correct ways to solve the problem; any answer given above is correct.
5	A	Each Loaf is cut into 8 parts Therefore the fraction of bread left : $= 1$ loaf $+ 1$ loaf $- \frac{5}{8}$ loaf $- \frac{7}{8}$ loaf $= 2 - \frac{12}{8}$ $= \frac{16 - 12}{8} = \frac{4}{8}$ $= \frac{1}{2}$ loaf

Question No.	Answer	Detailed Explanation
6	$\frac{5}{7}$	Using the key word altogether, we know we need to add. To add fractions, you first have to have the denominator the same. After that, add the numerators and keep the denominator the same. Finally, reduce if possible. $\frac{2}{7} + \frac{3}{7} = \frac{5}{7}$

Question 7

	$\frac{3}{5}$	$\frac{2}{3}$
There are two bags of candy. The first bag has $\frac{2}{5}$ of a bag, and the second has $\frac{1}{5}$ of a bag. What fraction of a bag of candy is there in all?	◯	
Izabel has a bag of marbles, but lets her brother have $\frac{1}{3}$ of them. What fraction represents how much she has left?		◯

We need to look at key words to figure out what operation we need to use. The first row is addition because of the words in all. To add fractions, you first have to have the denominator the same. After that, add the numerators and keep the denominator the same. Finally, reduce if possible. The second row is subtraction because she gave some to her brother, and we need to find out how much she has left. She started with a full bag, or $\frac{3}{3}$. Then she gave her brother $\frac{1}{3}$. So, $\frac{3}{3} - \frac{1}{3} = \frac{2}{3}$.

Question No.	Answer	Detailed Explanation
8	A & D	This is a problem on addition of fractions. First, express the fractions in terms of the least common denominator (LCD). Add the whole numbers and fractions separately and then put them together. LCD of 3 and 4 is 12. $\frac{1}{3} = \frac{1 \times 4}{3 \times 4} = \frac{4}{12}$; $\frac{1}{4} = \frac{1 \times 3}{4 \times 3} = \frac{3}{12}$

Question No.	Answer	Detailed Explanation
		Total pizza eaten $= 3\frac{1}{3} + 2\frac{1}{4} = 3\frac{4}{12} + 2\frac{3}{12}$ $= (3+2) + (\frac{4}{12} + \frac{3}{12}) = 5 + \frac{4+3}{12} = 5 + \frac{7}{12} = 5\frac{7}{12}$ To convert a mixed number to an improper fraction, multiply the denominator by the whole number, then add the numerator. This sum becomes the numerator of the improper fraction. The denominator of the improper fraction is the same as the denominator of the fraction part of the mixed number. $5\frac{7}{12} = \frac{(12\times5)+7}{4} = \frac{60+7}{12} = \frac{67}{12}$ pizzas Therefore, options (A) and (D) are correct.
9	$1\frac{13}{24}$	This is a problem on subtraction of fractions. We subtract the length of the ribbon used from the total length of ribbon. First, express the fractions in terms of the least common denominator (LCD). Subtract the whole numbers and fractions separately and then put them together. LCD of 12 and 8 is 24. $\frac{5}{12} = \frac{5\times2}{12\times2} = \frac{10}{24};\qquad \frac{7}{8} = \frac{7\times3}{8\times3} = \frac{21}{24}$ Length of the ribbon left $= 5\frac{5}{12} - 3\frac{7}{8} = 5\frac{10}{24} - 3\frac{21}{24}$ Since $\frac{10}{24} < \frac{21}{24}$, we regroup the first mixed number. $5\frac{10}{24} = 4 + \frac{24}{24} + \frac{10}{24} = 4 + \frac{24+10}{24} = 4 + \frac{34}{24}$ Length of the ribbon left $= 5\frac{10}{24} - 3\frac{21}{24} = 4\frac{34}{24} - 3\frac{21}{24}$ $= 4 + \frac{34}{24} - (3 + \frac{21}{24}) = (4-3) + (\frac{34}{24} - \frac{21}{24}) = 1 + \frac{34-21}{24}$ $= 1 + \frac{13}{24} = 1\frac{13}{24}$ m.

Lesson 7: Multiplying Fractions

Question No.	Answer	Detailed Explanation
1	C	Multiply $\frac{1}{2}$ x $\frac{6}{1}$. The answer is a fraction in which the numerator is larger than the denominator. This is called an improper fraction, which is always reduced to its lowest terms by dividing the denominator into the numerator: $\frac{6}{2}$. The quotient will be a whole number if there is no remainder. This is the answer.
2	B	The first answer is an improper fraction that, when divided into the numerator, has a remainder. Therefore, the answer will be a mixed fraction-a whole number and a fraction. The whole number represents how many times the denominator divides into the numerator. The remainder represents the numerator in the fraction and the denominator is the same denominator of the improper fraction in the original problem. Fraction part may be reduced to its simplest form. $6 \times \frac{1}{4} = (\frac{6}{1}) \times (\frac{1}{4}) = \frac{6 \times 1}{1 \times 4} = \frac{6}{4} = 1\frac{2}{4} = 1\frac{1}{2}$
3	D	One Whole + One Whole and later on multiply with $\frac{7}{10}$. $= 2 \times \frac{7}{10} = \frac{14}{10}$ Convert to mixed fraction $= 1\frac{4}{10}$
4	C	Convert the whole number into a fraction: $\frac{45}{1}$. Multiply the numerators together, then the denominators together. Reduce the answer to the lowest terms.
5	B & D	Multiply each pair of fractions by multiplying the numerators and multiplying the denominators. After that, simplify the fractions to see if they are equivalent to $\frac{1}{2}$. $\frac{2}{3} \times \frac{2}{3} = \frac{4}{9}$, $\frac{1}{2} \times \frac{10}{11} = \frac{10}{22} = \frac{5}{11}$, $\frac{1}{4} \times \frac{2}{6} = \frac{2}{24} = \frac{1}{12}$ are not equivalent to $\frac{1}{2}$. $\frac{2}{3} \times \frac{3}{4} = \frac{6}{12}$ simplifies to $\frac{1}{2}$. $\frac{3}{5} \times \frac{5}{6} = \frac{15}{30}$ simplifies to $\frac{1}{2}$.

Question No.	Answer	Detailed Explanation
7	$\frac{4}{15}$	Multiply each pair of fractions by multiplying the numerators and multiplying the denominators. After that, simplify the answer.
8		

	$\frac{8}{25}$	$\frac{1}{25}$	$\frac{1}{16}$
$\frac{4}{5} \times \frac{2}{5}$	◉		
$\frac{1}{4} \times \frac{1}{4}$			◉
$\frac{2}{5} \times \frac{1}{10}$		◉	

Multiply each pair of fractions by multiplying the numerators and multiplying the denominators. After that, simplify the answer if needed.

| 8 | A,C & D | $5 \times \frac{2}{7} = \frac{5}{1} \times \frac{2}{7} = \frac{5 \times 2}{1 \times 7} = \frac{10}{7}$. Therefore, option (A) is correct.

$\frac{10}{7} = 1\frac{3}{7}$. Therefore, option (C) is correct.

$5 \times \frac{2}{7}$ means adding $\frac{2}{7}$ five times. Therefore, option (D) is correct. |

Lesson 8: Multiplying Fractions by a Whole Number

Question No.	Answer	Detailed Explanation
1	D	To multiply a whole number by a unit fraction, multiply the whole number by 1 to get the numerator in the product. The denominator will not change.
2	B	Add up all of the fractional parts of meat. $\frac{2}{8} + \frac{2}{8} + \frac{2}{8} + \frac{2}{8} + \frac{2}{8} = \frac{10}{8}$ pounds. Then convert the improper fraction to a mixed number: $1\frac{2}{8}$ or $1\frac{1}{4}$ pounds.
3	A	When you multiply a whole number by a unit fraction (a fraction with 1 on top), the whole number becomes the numerator, and the denominator stays the same.
4	B	When you multiply a whole number by a unit fraction (a fraction with 1 on top), the whole number becomes the numerator, and the denominator stays the same.
5	A	All answers except for (A) will equal less than 1. $2 \times \frac{2}{3}$ equals $1\frac{1}{3}$.
6	8	We can solve this by guess and check if we want. To multiply a fraction by a whole number we multiply the numerator by the whole number, and leave the denominator alone. For example, we can pick and try 10. $\frac{1}{2} \times 10 = 5$, which is too big but we are close to 4. So we can try 8. $\frac{1}{2} \times 8 = 4$, so we know the missing number is 8.

7		$2\frac{2}{3}$	2	3
	$(\frac{1}{3}) \times 6$		◉	
	$(\frac{2}{3}) \times 4$	◯		
	$(\frac{3}{4}) \times 4$			◉

To multiply a fraction by a whole number we multiply the numerator by the whole number, and leave the denominator alone. Then simplify into a mixed number or whole number.

Question No.	Answer	Detailed Explanation
8	A, C, D	When we multiply a whole number n by a fraction $\frac{a}{b}$, the product is equal to $\frac{n \times a}{b}$. $3 \times \frac{7}{6} = \frac{3 \times 7}{6} = \frac{21}{6}$. Therefore, option (A) is correct. $\frac{21}{6} = 3\frac{3}{6}$; but $\frac{3}{6} = \frac{(3 \div 3)}{(6 \div 3)} = \frac{1}{2}$ Therefore, $\frac{21}{6} = 3\frac{1}{2}$. Therefore, option (C) is correct. $3 \times \frac{7}{6}$ means adding $\frac{7}{6}$ three times. Therefore, option (D) is correct.

Lesson 9: Multiplying Fractions in Word Problems

Question No.	Answer	Detailed Explanation
1	B	Students may use the multiplication equation $9 \times \frac{2}{3} = \frac{18}{3}$, which simplifies to 6 cups of mint M&Ms. Students may use the repeated addition equation $\frac{2}{3} + \frac{2}{3} + \frac{2}{3} + \frac{2}{3} + \frac{2}{3} + \frac{2}{3} + \frac{2}{3} + \frac{2}{3} + \frac{2}{3} = \frac{18}{3}$ or 6 cups of mint M&Ms.
2	A	Students could have used: $9 \times \frac{1}{2} = \frac{9}{2}$ or $4\frac{1}{2}$ cups of coconut M&Ms. Or, students could have used: $\frac{1}{2} + \frac{1}{2} + \frac{1}{2} + \frac{1}{2} + \frac{1}{2} + \frac{1}{2} + \frac{1}{2} + \frac{1}{2} + \frac{1}{2} = \frac{9}{2}$ or $4\frac{1}{2}$ cups of coconut M&Ms.
3	B	Student may draw a model that shows that $\frac{2}{3} + \frac{1}{2} > 1$. Student may state that $\frac{2}{3}$ is more than $\frac{1}{2}$. If $\frac{1}{2} + \frac{1}{2} = 1$, then $\frac{2}{3} + \frac{1}{2} > 1$. Students may convert fractions to common denominators and add to find the total is $\frac{7}{6}$ or $1\frac{1}{6}$ cups of M&Ms.
4	B	$(\frac{3}{4}) \times 7 = 5\frac{1}{4}$ miles. $\frac{3}{4} + \frac{3}{4} + \frac{3}{4} + \frac{3}{4} + \frac{3}{4} + \frac{3}{4} + \frac{3}{4} = 5\frac{1}{4}$ miles.
5	A	$(\frac{3}{5}) \times 4 = 2\frac{2}{5}$, or $\frac{3}{5} + \frac{3}{5} + \frac{3}{5} + \frac{3}{5} = 2\frac{2}{5}$
6	2	We need to figure out how much is 1/3, and then subtract it from what he has. 1/3 x 3 = 1 cup. If Tim used 1 cup, he has 2 cups left

7

	$40	$60
The dinner for a large family costs $80. Mr. Smith has a 1/2 off coupon, so what will the final price be?	◉	
The movie tickets cost $90 but I have a coupon for 1/3 off. How much will I spend?		◉

First row: We need to figure out how much is $\frac{1}{2}$ of the 80, and then subtract it from the 80. $\frac{1}{2} \times 80 = 40$. $80 - 40 = 40$, so they spent $40.

Second row: We need to figure out how much is $\frac{1}{3}$ of the 90, and then subtract it from the 90. $\frac{1}{3} \times 90 = 30$. $90 - 30 = 60$, so they spent $60.

Question No.	Answer	Detailed Explanation
8	D	Multiply 8 by $\frac{2}{5}$ to get the number of tables Jose can paint. $8 \times \frac{2}{5} = \frac{8 \times 2}{5} = \frac{16}{5} = 3\frac{1}{5}$. Jose can paint $3\frac{1}{5}$ tables in 15 minutes.

Lesson 10: 10 to 100 Equivalent Fractions

Question No.	Answer	Detailed Explanation
1	A	6 dimes is 60 pennies, so 6 dimes and 3 pennies is 63 pennies, which is $\frac{63}{100}$=0.63 of a dollar
2	A	<table><tr><td>thousands</td><td>hundreds</td><td>tens</td><td>units</td><td>• decimal</td><td>tenths</td><td>hundredths</td><td>thousandths</td><td>ten thousandths</td></tr><tr><td></td><td></td><td></td><td></td><td></td><td>1</td><td>4</td><td></td><td></td></tr></table> 0.14 is read as 14 hundredths $\frac{14}{100}$.
3	C	<table><tr><td>thousands</td><td>hundreds</td><td>tens</td><td>units</td><td>• decimal</td><td>tenths</td><td>hundredths</td><td>thousandths</td><td>ten thousandths</td></tr><tr><td></td><td></td><td></td><td></td><td></td><td>5</td><td>2</td><td></td><td></td></tr></table> 0.52 is read as 52 hundredths $\frac{52}{100}$
4	A	<table><tr><td>thousands</td><td>hundreds</td><td>tens</td><td>units</td><td>• decimal</td><td>tenths</td><td>hundredths</td><td>thousandths</td><td>ten thousandths</td></tr><tr><td></td><td></td><td></td><td></td><td></td><td>2</td><td>5</td><td></td><td></td></tr></table> 0.25 is read as twenty five hundredths $\frac{25}{100}$.
5	D	The correct answer is D. 14 hundredths = 10 hundredths + 4 hundredths

Question No.	Answer		Detailed Explanation

| 6 | C |

Since there is 4 hundredths on the right hand side, we need to split 14 hundredths into 10 hundredths plus 4 hundredths. 10 hundredths $= \frac{10}{100} = \frac{1}{10}$. Or it can be done as follows :
14 hundredths $= \frac{14}{100} = \frac{10+4}{100} = \frac{10}{100} + \frac{4}{100} = \frac{1}{10} + \frac{4}{100} = 1$ tenths $+ 4$ hundredths.

| 7 | B |

14 hundredths = 10 hundredths + 4 hundredths = 1 tenths + 4 hundredths = 1 tenths + 3 hundredths + 1 hundredths. Therefore Option (B) is the correct answer.

Or $\frac{14}{100} = \frac{10}{100} + \frac{4}{100} = \frac{1}{10} + \frac{4}{100} = \frac{1}{10} + \frac{(3+1)}{100}$

$= \frac{1}{10} + \frac{3}{100} + \frac{1}{100}$

| 8 | A |

80 hundredths = 0.80. Zero at the end of the decimal number has no value. Therefore 0.80 = 0.8 = 8 tenths.

or 80/100 = 8/10 = 8 tenths.

Question No.	Answer	Detailed Explanation
9	C	Students must realize that $\frac{2}{10}$ is equivalent to $\frac{20}{100}$ and then add $\frac{20}{100} + \frac{41}{100}$

10			

	$\frac{10}{100}$	$\frac{60}{100}$	$\frac{90}{100}$
$\frac{9}{10}$			◯
$\frac{1}{10}$	◯		
$\frac{6}{10}$		◯	

To find equivalent fractions, you have to multiply the numerator and denominator by the same number. In this case, we want to get the denominator to 100, so we have to multiply each by 10

| 11. | C | To find equivalent fractions, you have to multiply the numerator and denominator by the same number. In this case, we want to get the denominator to 100, so we have to multiply each by 10. 2 x 10 = 20, 10 x 10 = 100. So we have $\frac{20}{100}$ |

| 12 | $\frac{40}{100}$ | To find equivalent fractions, you have to multiply the numerator and denominator by the same number. In this case, we want to get the denominator to 100, so we have to multiply each by 10. 4 x 10 = 40, 10 x 10 = 100. So we have $\frac{40}{100}$ |

Question No.	Answer	Detailed Explanation
13	A & C	Write the equivalent fraction of $\frac{3}{10}$ with 100 as the denominator.

$$\frac{3}{10} = \frac{3 \times 10}{10 \times 10} = \frac{30}{100}$$

Subtract the whole numbers and the fractions separately and then put them together.

Since $\frac{5}{100} < \frac{30}{100}$, we regroup the first mixed number.

$$8\frac{5}{100} = 7 + \frac{100}{100} + \frac{5}{100} = 7 + \frac{100+5}{100} = 7 + \frac{105}{100} = 7\frac{105}{100}$$

$$8\frac{5}{100} - 5\frac{3}{10} = 7\frac{105}{100} - 5\frac{30}{100} = 7 + \frac{105}{100} - (5 + \frac{30}{100})$$

$$= (7-5) + (\frac{105}{100} - \frac{30}{100}) = 2 + \frac{105-30}{100} = 2 + \frac{75}{100} = 2\frac{75}{100}$$

$$\frac{75}{100} = \frac{(75 \div 25)}{(100 \div 25)} = \frac{3}{4} \qquad So, \ 2\frac{75}{100} = 2\frac{3}{4}$$

Therefore, options (A) and (C) are correct.

Lesson 11: Convert Fractions to Decimals

Question No.	Answer	Detailed Explanation
1	C	Point A is located near $\frac{1}{4}$ of 1 on the number line. In order to represent numbers that are less than 1, a fraction or decimal is used. To convert the fraction to a decimal, divide the numerator by the denominator, adding zeros until there is no remainder. A decimal point is plotted after the last digit in the dividend. Another decimal point is plotted directly above that one in the answer. The decimal equivalent of $\frac{1}{4}$ is 0.25.
2	A	This fraction would be read "one hundred forty-eight thousandths." As a decimal, that number is written 0.148.
3	C	Point D is halfway between -1 and -2. Halfway between -1 and -2 is -1.5.
4	A	Convert the fraction part of this problem into a decimal by dividing the denominator into the numerator: 1 divided by 4. Put the whole number part of this problem in front of the decimal to complete the answer.
5	B	This fraction would be read "three thousandths." As a decimal, it is written 0.003.
6	D	The fraction this model represents is $\frac{27}{100}$: there are 27 cells shaded and 100 cells altogether. Twenty-seven hundredths is written as 0.27 in decimal form.
7	C	The last digit in this decimal is in the thousandths place. 44 thousands as a fraction would be written $\frac{44}{1000}$.
8	B	Divide the denominator into the numerator to change fraction to a decimal. Put the whole number in front of the decimal.
9	C	The last digit in the decimal is in the thousandths place. So, 0.193 would be written as $\frac{193}{1000}$ in fraction form.
10	D	Convert the decimals into fractions to be added. Both of these decimals end in the thousandths place. So the addends are $\frac{300}{1000}$ and $\frac{249}{1000}$.

	0.50	0.25	0.10
$\frac{1}{2}$	◯		
$\frac{1}{4}$		◯	
$\frac{1}{10}$			◯

To convert fractions to decimals, change the fraction to have a denominator of 100. For example, in the first row, we would need to multiply 2 by 50 to get 100. So we have to do that to the numerator as well. $\frac{1}{2} \times \frac{50}{50} = \frac{50}{100}$

The numerator is the number after the decimal, with the tens digit in the tenths place, and the ones digit in the hundredths place because decimals are base 100, and percent means per 100. So $\frac{50}{100}$ is equivalent to 0.50, or simply 0.5. You can drop the zeros on the far right, but if they are between other numbers, they have to stay because they are placeholders

$\frac{25}{100}$ converts to 0.25 and $\frac{1}{10}$ converts to 0.10.

Question No.	Answer	Detailed Explanation
12	B & E	To convert fractions to decimals, change the fraction to have a denominator of 100. For example, in the first row, we would need to multiply 2 by 50 to get 100. So we have to do that to the numerator as well. $\frac{1}{2} \times \frac{50}{50} = \frac{50}{100}$

The numerator is the number after the decimal, with the tens digit in the tenths place, and the ones digit in the hundredths place. So $\frac{1}{2}$ is equivalent to 0.50, or simply 0.5. You can drop the zeros on the far right, but if they are between other numbers, they have to stay because they are placeholders. |

Question No.	Answer	Detailed Explanation
		$\frac{1}{4} = \frac{(1 \times 25)}{(4 \times 25)} = \frac{25}{100} = 25$ hundredths $= 0.25$. Therefore, $\frac{1}{4}$ and $\frac{25}{4}$ are equivalent to 0.25.
		$\frac{9}{10} = \frac{(9 \times 10)}{(10 \times 10)} = \frac{90}{100} = 90$ hundredths $= 0.90$. Therefore $\frac{9}{10}$ is not equivalent to 0.25.
		$\frac{3}{5} = \frac{(3 \times 20)}{(5 \times 20)} = \frac{60}{100} = 60$ hundredths $= 0.60$. Therefore $\frac{3}{5}$ is not equivalent to 0.25.
13	0.05	$\frac{1}{20} = \frac{1 \times 10}{20 \times 10}$ $\frac{10}{2 \times 10 \times 10} = \frac{5}{100}$ 5 Hundredth $= 0.05$
14	A,C & D	$\frac{6}{10}$ is read as 6 tenths. Here, the denominator is 10. Therefore, we can directly write the numerator putting the decimal in the correct spot (one space from the right-hand side for every zero in the denominator); 0.6 Therefore, option (A) is correct. $\frac{6}{10} = \frac{6 \times 10}{10 \times 10} = \frac{60}{100}$ Multiply and divide 0.06 by 100 to remove the decimal point. $0.06 = \frac{0.06 \times 100}{100} = \frac{6.00}{100} = \frac{6}{100}$. $\frac{6}{100}$ is not equal to $\frac{6}{10}$. Therefore, option (B) is wrong. Therefore, option (C) is correct. 0.60 is the same as 0.6 (the zero at the end of 0.60 has no value.) Therefore, option (D) is correct.

Lesson 12: Compare Decimals

Question No.	Answer	Detailed Explanation
1	A	The point is plotted at about $1\frac{3}{4}$. Convert this mixed fraction to a decimal. $1\frac{3}{4} = 1.75$
2	A	0.05 is 5 hundredths. 0.50 is 50 hundredths. 0.05 < 0.50
3	B	All of the dinner prices consist of decimals that have numbers to the left of the decimal point. Those numbers are whole numbers (dollars). To compare, simply look at the dollar amounts. The cents do not matter as long as the dollar amounts are all different. Note: The cost of the chicken dinner is $25.79. (30.79 - 5.00 = 25.79)
4	B	Zeros can be added after the last digit in a decimal number without changing its value. Therefore, 0.2 = 0.200.
5	C	All of these decimals have the same digit in the tenths place (a 0). Compare the digits in the hundredths place. The number that does not have a zero in the hundredths place has the greatest value.
6	B	Compare the digits in the ones place. The number with a non-zero digit in the ones place has the greater value. 1.954 > 0.1954
7	C	Numbers in the tenths place are greater than numbers in the hundredths place, numbers in the hundredths place are greater than numbers in the thousandths place, numbers in the thousandths place are greater than numbers in the ten thousandths place, and whole numbers are the greatest of them. Use the placement of the 4 to compare and order the numbers.
8	B	The zeros after the last digit in each number do not change its value. 1.10 = 1.1000.
9	B	Add $65 to the weekday price. $225 + 65 = $290
10	B	There is a pattern here that is related to the size of the vehicle and an increase of $5,000.00. Each vehicle in the table costs $5,000 more than the vehicle listed above it.

Question No.	Answer	Detailed Explanation
11	0.93	To find the greatest decimal, look at the tenths digit (one after the decimal). The biggest number is the greatest. If there are two numbers with the same digit in the tenths spot, move to the hundredths.

Question 12

	<	>	=
0.25 _ 0.39	◯		
0.89 _ 0.890			◯
0.12 _ 0.21	◯		
0.29 _ 0.28		◯	

To compare decimals, look at each digit individually. Looking at the tenths digit, if they are different, the one with the larger number in the tenths place is the larger number. If the tenths are the same, move one digit to the right, and repeat that process with the hundredths place. If there is a zero in the last digit like in row two, those numbers are still equivalent, because you can drop the zero.

Question 13

0.28,0.39,0.42,0.04,0.37

To compare decimals, look at each digit individually. Looking at the tenths digit, if they are different, the one with the larger number in the tenths place is the larger number. If the tenths are the same, move one digit to the right, and repeat that process with the hundredths place. So we choose the numbers that have the tenths digit being smaller than 4, or if the tenths digit is 4, the hundredths place has to be smaller than 5. So we see the correct answers are 0.28, 0.39, 0.42, 0.04, and 0.37.

Question 14

In the first area model, 46 out of 100 cells are shaded. It represents $\frac{46}{100} = 0.46$. X = 0.46

In the second area model, 45 out of 100 cells are shaded. It represents $\frac{45}{100} = 0.45$. Y = 0.45

Therefore, X > Y

Question No.	Answer	Detailed Explanation
15	>,>,=	(1) 8 ones and 5 hundredths = $(8 \times 1) + 5 \times (\frac{1}{100}) = 8 + \frac{5}{100} = 8 + 0.05 = 8.05$. $8.05 < 8.5$ (because 8.05 has 0 in the tenths place and 8.5 has 5 in the tenths place)
		(2) $\frac{56}{10} = 5.6$; $7\frac{56}{10} = 7 + \frac{56}{10} = 7 + 5.6 = 12.6 > 7.56$ (because 12.6 has 1 in the tens place and 7.56 has 0 in the tens place)
		(3) 5 tenths = $5 \times (\frac{1}{10}) = \frac{5}{10} = 0.5$; 25 hundredths = $25 \times (\frac{1}{100}) = \frac{25}{100} = 0.25$; $0.5 > 0.25$ (because 0.5 has 5 in the tenths place and 0.25 has 2 in the tenths place)
		(4) 65 tenths and 2 hundredths = $65 \times (\frac{1}{10}) + 2 \times (\frac{1}{100}) = \frac{65}{10} + \frac{2}{100} = 6.5 + 0.02 = 6.52$
		Therefore, 6.52 is equal to 65 tenths and 2 hundredths.

LumosLearning.com

Chapter 4: Measurement and Data

Lesson 1: Units of Measurement

1. **What customary unit should be used to measure the weight of the table shown in the picture below?**

Ⓐ pounds
Ⓑ inches
Ⓒ kilograms
Ⓓ tons

2. **Which of the following is an appropriate customary unit to measure the weight of a small bird?**

Ⓐ grams
Ⓑ ounces
Ⓒ pounds
Ⓓ units

3. **Complete the following statement:**
   A horse might weigh _____.

Ⓐ about 500 pounds
Ⓑ about 12 pounds
Ⓒ about a gallon
Ⓓ about 200 ounces

4. **Choose the appropriate customary unit to measure the length of a road.**

Ⓐ Yard
Ⓑ Meter
Ⓒ Kilometer
Ⓓ Mile

5. **Choose the appropriate unit to measure the height of a tall tree.**

   Ⓐ miles
   Ⓑ yards
   Ⓒ centimeters
   Ⓓ gallons

6. **Which of the following is not a customary unit?**

   Ⓐ Kilograms
   Ⓑ Yards
   Ⓒ Pounds
   Ⓓ Miles

7. **Which of the following is a customary unit that can be used to measure the volume of a liquid?**

   Ⓐ yards
   Ⓑ fluid ounces
   Ⓒ milliliters
   Ⓓ pounds

8. **Which is more, 18 teaspoons or 2 fluid ounces?**

   Ⓐ 2 fluid ounces
   Ⓑ They are equal.
   Ⓒ 18 teaspoons

9. **A math textbook might weigh _____ .**

   Ⓐ 125 ounces
   Ⓑ 25 ounces
   Ⓒ 200 ounces
   Ⓓ 2 ounces

10. **To make the medicine seem to taste better, Mom told Bonita she had to take tablespoons instead of teaspoons. How many tablespoons of medicine should Bonita take if the dose is 3 teaspoons?**

    Ⓐ tablespoons
    Ⓑ 1 tablespoon
    Ⓒ 6 tablespoons
    Ⓓ 1/2 tablespoon

11. How many kilometers are there in 4,000 meters? Write your answer in the box below

12. Match each row with the correct conversion

	4	5	6
360 min = ___ hr	○	○	○
500 cm = ___ m	○	○	○
240 sec = ___ min	○	○	○

13. Fill in the missing values in the table based on the conversions shown in the header.

km	m	cm
6.5	6,500	650,000
8.2		
	7,300	
		825,000

14. Complete the following. 44 pints = ? and select all the correct answers.

Ⓐ 88 cups
Ⓑ 22 cups
Ⓒ 88 quarts
Ⓓ 22 quarts
Ⓔ 5.5 gallons
Ⓕ 11 gallons

CHAPTER 4 →Lesson 2: Measurement Problems

1. Arthur wants to arrive at soccer practice at 5:30 PM. He knows it takes him 42 minutes to walk to practice from his house. Estimate the time Arthur should leave his house to go to practice?

   Ⓐ 5:00 PM
   Ⓑ 4:45 PM
   Ⓒ 4:30 PM
   Ⓓ 3:45 PM

2. A baseball game began at 7:05 PM and lasted for 2 hours and 38 minutes. At what time did the game end?

   Ⓐ 9:43 PM
   Ⓑ 10:33 PM
   Ⓒ 9:38 PM
   Ⓓ 9:33 PM

3. 4 feet and 5 inches is the same as:

   Ⓐ 48 inches
   Ⓑ 53 inches
   Ⓒ 41 inches
   Ⓓ 65 inches

4. Amir bought two cowboy hats for $47, a pair of cowboy boots for $150, and a leather belt for $32. The tax was $13.74. He gave the cashier $300. How much change does she owe him?

   Ⓐ $242.74
   Ⓑ $13.74
   Ⓒ $57.26
   Ⓓ $257.26

5. Harriet needed $\frac{1}{2}$ cup of milk for the white sauce, but she could only find her table-spoon to measure with. How many tablespoons of milk will she need?

   Ⓐ 4 tablespoons
   Ⓑ 6 tablespoons
   Ⓒ 8 tablespoons
   Ⓓ 10 tablespoons

6. Use a comparison symbol to complete the following statement:
   32 ounces ___ 1 pound

   (A) <
   (B) =
   (C) >

7. Rachel's gymnastics lessons lasted for 1 year. Sharon's lessons lasted for 9 months. Yolanda's lessons lasted for 23 months. How much time did the girls spend on lessons altogether?

   (A) 4 years, 0 months
   (B) 3 years, 0 months
   (C) 3 years, 8 months
   (D) 4 years, 8 months

8. To make some of the best cookies, mix 1 cup of butter, 2 cups of sugar, $2\frac{1}{2}$ cups of flour, $2\frac{1}{2}$ teaspoons of vanilla extract, $\frac{1}{2}$ teaspoon of baking soda, and $\frac{3}{4}$ cups of chocolate chips. Which comparison symbols would complete the following statements?

   amount of flour ___ amount of butter
   amount of butter ___ amount of sugar

   (A) >; =
   (B) <; >
   (C) >; <
   (D) <; =

9. The bookstore is selling paperback books for $3.25 each. How much would 4 paperback books cost?

   (A) $12.00
   (B) $13.00
   (C) $13.50
   (D) $12.50

10. If Cindy bought 3 DVDs and 2 nacho kits, how much would she pay for all items before taxes? Use the table below to answer the question:

Item	Unit Price
CDs	$10.99
DVDs	$24.99
Cordless Phone	$30.00
Flash Drives	$9.99
Nacho Kits	$6.99

Ⓐ $31.98
Ⓑ $74.97
Ⓒ $84.95
Ⓓ $88.95

11. Sam was measuring out 6 cups of milk, but his measuring cup only measured 2 cups. How many times does he fill the measuring cup? Write your answer in the box below.

12. Select the correct answer for each row.

	$2\frac{1}{4}$	$4\frac{1}{4}$
I need 5 cups of flour for my recipe. I need ¾ cup more. How much flour did I start with?	○	○
I want to make a tug of war rope. It has to be 16 feet long. I have 18 feet 9 inches of rope. How much will I have to cut off? Express your answer as a mixed number in feet.	○	○

13. Complete the table.

On Thursday, the restaurant used 2 cups less sugar than they used on Friday. On Friday, they used 22 cups. On Saturday, they used 3 more than on Friday. What was the total amount of sugar used over those three days?	
Anna was cutting wood. She cut 5 cords of wood on Monday, 1 less than that on Tuesday, and 3 more than that on Wednesday (compared to Monday). How much wood did she cut in all?	

14. Greg rode his bicycle for the past 5 days. He covered $2\frac{1}{4}$ km every day. How much distance did he cover in 5 days? Express your answer.

Ⓐ 1,125 meters
Ⓑ 12,250 meters
Ⓒ 1,125 meters
Ⓓ 11,250 meters

CHAPTER 4 →Lesson 3: Perimeter & Area

1. **A rectangular room measures 10 feet long and 13 feet wide. How could you find out the area of this room?**

   Ⓐ Add 10 and 13, then double the results
   Ⓑ Multiply 10 by 13
   Ⓒ Add 10 and 13
   Ⓓ None of the above

2. **A rectangle has a perimeter of 30 inches. Which of the following could be the dimensions of the rectangle?**

   Ⓐ 10 inches long and 5 inches wide
   Ⓑ 6 inches long and 5 inches wide
   Ⓒ 10 inches long and 3 inches wide
   Ⓓ 15 inches long and 15 inches wide

3. **Which of these expressions could be used to find the perimeter of the figure below?**

   48 feet

   ```
   ┌─────────────────────┐
   │                     │
   │                     │
   │      Figure A       │   36 feet
   │                     │
   │                     │
   └─────────────────────┘
   ```

   Ⓐ 48 + 36 + 2
   Ⓑ 48 x 36
   Ⓒ 2 x (48 + 36)
   Ⓓ 48 + 36

4. **A chalkboard is 72 inches long and 30 inches wide. What is its perimeter?**

   Ⓐ 204 inches
   Ⓑ 2,160 inches
   Ⓒ 102 inches
   Ⓓ 2,100 inches

5. Which of the following statements is true?

Figure A

12 feet

8 feet

Figure B

10 feet

10 feet

Ⓐ The two shapes have the same perimeter.
Ⓑ The two shapes have the same area.
Ⓒ The two figures are congruent.
Ⓓ Figure A has a greater area than Figure B.

6. If a square has a perimeter of 100 units, how long is each of its sides?

Ⓐ 10 units
Ⓑ 20 units
Ⓒ 25 units
Ⓓ Not enough information is given.

7. Find the perimeter of Shape C.

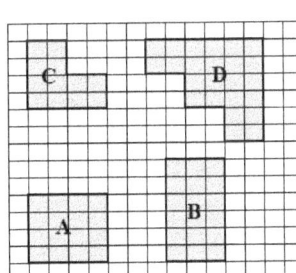

☐ = 1 square unit

Ⓐ 8 units
Ⓑ 12 units
Ⓒ 14 units
Ⓓ 16 units

8. **Find the area of Shape C.**

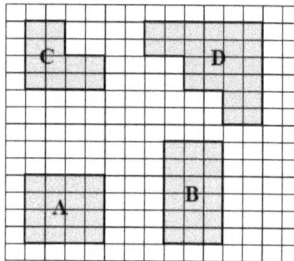

$\square$ = 1 square unit

Ⓐ 8 square units
Ⓑ 12 square units
Ⓒ 14 square units
Ⓓ 16 square units

9. **What is the perimeter of this shape?**

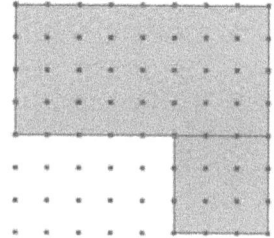

Ⓐ 24 units
Ⓑ 28 units
Ⓒ 30 units
Ⓓ 34 units

10. **A rectangle has an area of 48 square units and a perimeter of 32 units. What are its dimensions?**

Ⓐ 6 units by 8 units
Ⓑ 12 units by 4 units
Ⓒ 16 units by 3 units
Ⓓ All of the above are possible.

11. What is the perimeter of the following 6 cm by 4 cm rectangle? Write the answer in the box given below

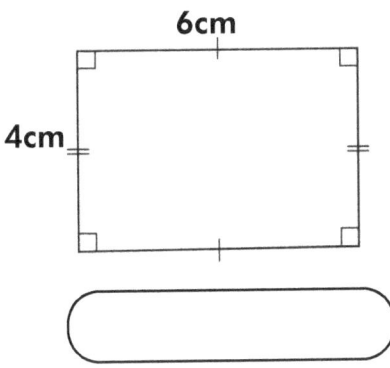

12. Select the correct perimeter and area values for the figure shown below.

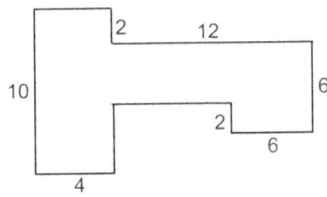

	100	56	102
Perimeter: ___ units	○	○	○
Area: ___ units squared	○	○	○

13. Fill in the correct perimeter and area for the following triangle.

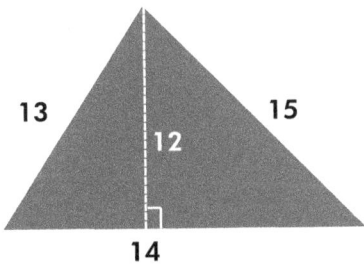

Area	
Perimeter	

14. If the area of a square paper is 144 sq. feet, what is the length of any side of the paper? Circle the correct answer.

Ⓐ 12 feet
Ⓑ 36 feet
Ⓒ 18 feet
Ⓓ 6 feet

CHAPTER 4 →Lesson 4: Representing and Interpreting Data

1. The students in Mrs. Riley's class were asked how many cousins they have. The results are shown in the line plot. Use the information shown in the line plot to respond to the following.
   How many of the students have no cousins?

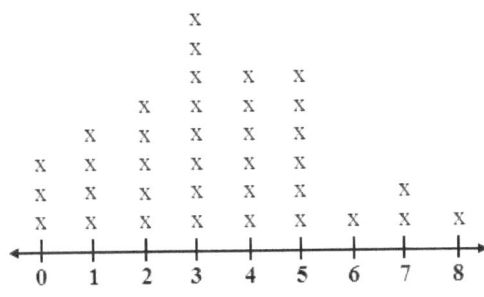

   Ⓐ 0 students
   Ⓑ 1 student
   Ⓒ 2 students
   Ⓓ 3 students

2. The students in Mrs. Riley's class were asked how many cousins they have. The results are shown in the line plot. Use the information shown in the line plot to respond to the following.
   How many of the students have exactly 4 cousins?

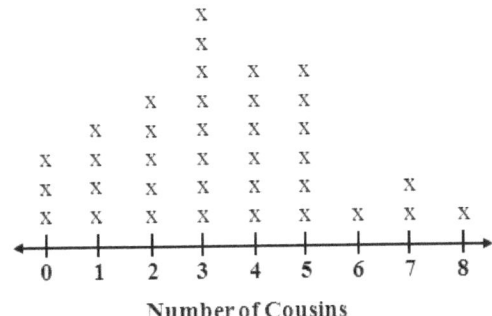

   Ⓐ 1 student
   Ⓑ 5 students
   Ⓒ 6 students
   Ⓓ 7 students

3. According to this graph, which are the 2 most favorite foods people enjoy at a carnival?

Favorite Carnival Foods

caramel apples	elephant ears	corn dogs	cotton candy	french fries	candy apples	funnel cakes
x			x			
x			x		x	x
x			x		x	x
x			x		x	x
x		x	x		x	x
x		x	x		x	x
x	x	x	x		x	x
x	x	x	x		x	x
x	x	x	x	x	x	x
x	x	x	x	x	x	x
x	x	x	x	x	x	x

Ⓐ candy apples and funnel cake
Ⓑ caramel apples and cotton candy
Ⓒ cotton candy and funnel cake
Ⓓ caramel apples and candy apples

4. How many more families prepare the night before the picnic than prepare right before leaving for the picnic?

Families That Pack the Night Before the Picnic	Families That Rise Early in the Morning to Pack for the Picnic	Families That Pack Right Before They Leave for the Picnic
x		
x	x	
x	x	x
x	x	x

(Each X represents 2 families.)

Ⓐ 2
Ⓑ 4
Ⓒ 1
Ⓓ 8

5. The school nurse kept track of how many students visited her clinic during a week. She plotted the results on a line graph. Use the line graph to respond to the following. What trend should the nurse notice?

Nurse's Clinic Visits

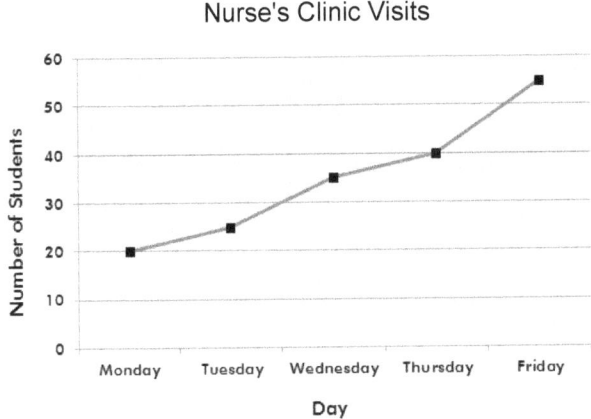

Ⓐ The number of students visiting her office increased throughout the week.
Ⓑ The number of students visiting her office decreased throughout the week.
Ⓒ The number of students visiting her office remained constant throughout the week.
Ⓓ There is no apparent trend.

6. The school nurse kept track of how many students visited her clinic during a week. She plotted the results on a line graph. Use the line graph to respond to the following. How many more students visited the nurse on Friday than on Monday?

Nurse's Clinic Visits

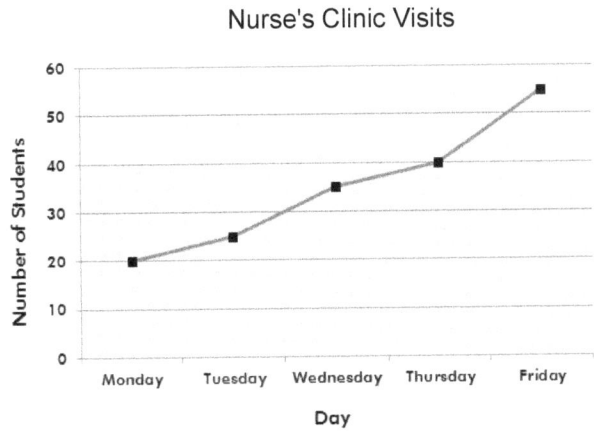

Ⓐ 20 students
Ⓑ 35 students
Ⓒ 45 students
Ⓓ 55 students

7. The school nurse kept track of how many students visited her office during a week. She plotted the results on a line graph. Use the line graph to respond to the following. How many students visited the nurse on Wednesday?

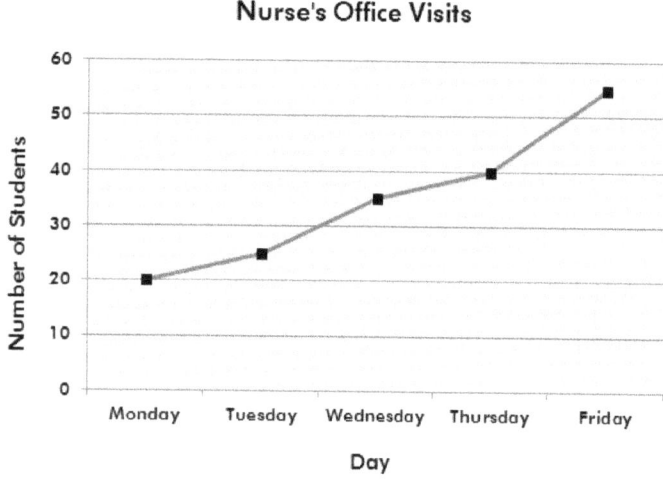

Nurse's Office Visits

Ⓐ 30 students
Ⓑ 35 students
Ⓒ 40 students
Ⓓ 45 students

8. The fourth grade chorus is selling candy boxes to raise money for a trip to the water park. The pictograph below shows how many candy boxes they sold during the first four weeks of the sale. Use the information shown in the graph to respond to the following.
The chorus needed to sell 150 boxes of candy to pay for the trip. Did they sell enough boxes?

Candy Boxes Sold

Week 1	☐ ☐ ☐ ☐
Week 2	☐ ☐ ☐ ◣
Week 3	☐ ☐ ◣
Week 4	☐ ☐ ☐ ☐ ☐ ◣

Key : ☐ = 10 boxes
◣ = 5 boxes

Ⓐ Yes, they sold more than enough boxes.
Ⓑ Yes, they sold exactly 150 boxes.
Ⓒ No, they needed to sell 5 more boxes.
Ⓓ No, they needed to sell 10 more boxes.

9. The fourth grade chorus is selling candy boxes to raise money for a trip to the water park. The pictograph below shows how many candy boxes they sold during the first four weeks of the sale. Use the information shown in the graph to respond to the following.
    How many more candy boxes were sold during the fourth week than during the first week?

    Candy Boxes Sold

    | Week 1 | □ □ □ □ |
    | Week 2 | □ □ □ ◿ |
    | Week 3 | □ □ ◿ |
    | Week 4 | □ □ □ □ □ ◿ |

    Key : □ = 10 boxes
          ◿ = 5 boxes

    Ⓐ 25 boxes
    Ⓑ 15 boxes
    Ⓒ 5 boxes
    Ⓓ None of the above

10. The fourth grade chorus is selling candy boxes to raise money for a trip to the water park. The pictograph below shows how many candy boxes they sold during the first four weeks of the sale. Use the information shown in the graph to respond to the following.
    How many candy boxes were sold during the first two weeks of the sale?

    Candy Boxes Sold

    | Week 1 | □ □ □ □ |
    | Week 2 | □ □ □ ◿ |
    | Week 3 | □ □ ◿ |
    | Week 4 | □ □ □ □ □ ◿ |

    Key : □ = 10 boxes
          ◿ = 5 boxes

    Ⓐ $7\frac{1}{2}$ boxes
    Ⓑ 40 boxes
    Ⓒ 35 boxes
    Ⓓ 75 boxes

11. Mr. Green's class spent 5 weeks collecting cans as part of a recycling project. The bar graph shows how many cans they collected each week. Use the graph to respond to the following question.

How many cans were collected during the second week?

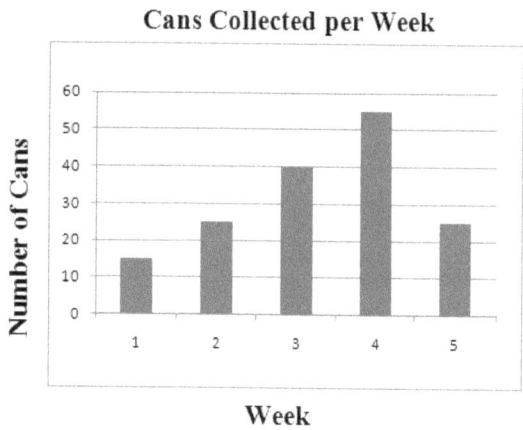

Ⓐ 15 cans
Ⓑ 20 cans
Ⓒ 25 cans
Ⓓ 35 cans

12. Mr. Green's class spent 5 weeks collecting cans as part of a recycling project. The bar graph shows how many cans they collected each week. Use the graph to respond to the following question.

How many more cans were collected during the third week than during the second week?

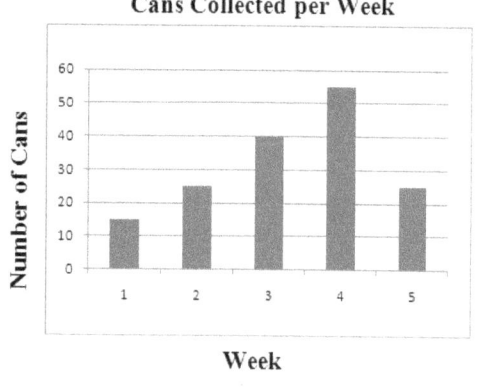

Ⓐ 5 cans
Ⓑ 15 cans
Ⓒ 25 cans
Ⓓ 40 cans

13. Mr. Green's class spent 5 weeks collecting cans as part of a recycling project. The bar graph shows how many cans they collected each week. Use the graph to respond to the following question.
How many cans were collected during the fourth week?

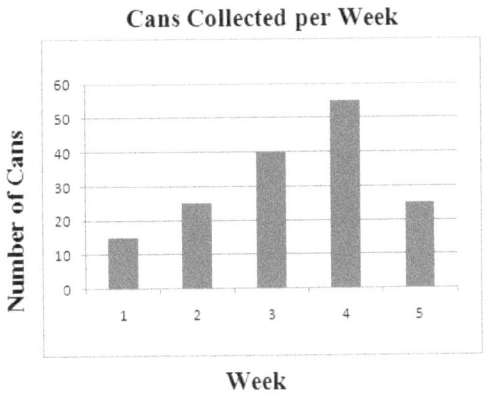

Ⓐ 65 cans
Ⓑ 55 cans
Ⓒ 50 cans
Ⓓ 40 cans

14. Mr. Green's class spent 5 weeks collecting cans as part of a recycling project. The bar graph shows how many cans they collected each week. Use the graph to respond to the following question.
During which week were the fewest cans collected?

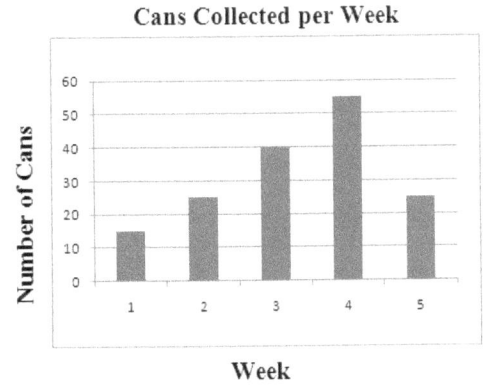

Ⓐ Week 1
Ⓑ Week 2
Ⓒ Week 3
Ⓓ Week 4

15. Mr. Green's class spent 5 weeks collecting cans as part of a recycling project. The bar graph shows how many cans they collected each week. Use the graph to respond to the following question.

    During which two weeks were the same number of cans collected?

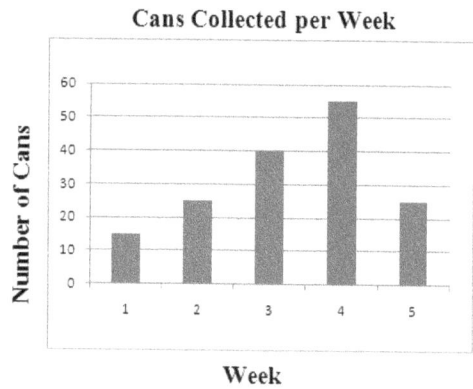

Ⓐ Weeks 1 and 2
Ⓑ Weeks 2 and 4
Ⓒ Weeks 3 and 5
Ⓓ Weeks 2 and 5

16. The students in the third grade were surveyed to find out their favorite seasons. The results are shown in the tally table. Use the tally table to respond to the following.

    Which season was chosen as the favorite of the most students?

Our Favorite Seasons

Winter	⊞⊞ ⊞⊞ ⊞⊞ II
Spring	⊞⊞ II
Summer	⊞⊞ ⊞⊞ ⊞⊞ III
Fall	III

Ⓐ Winter
Ⓑ Spring
Ⓒ Summer
Ⓓ Fall

17. The students in the third grade were surveyed to find out their favorite seasons. The results are shown in the tally table. Use the tally table to respond to the following. How many students chose winter as their favorite season?

Our Favorite Seasons

Winter															
Spring							
Summer																
Fall				

Ⓐ 12 students
Ⓑ 16 students
Ⓒ 17 students
Ⓓ 22 students

18. The students in the third grade were surveyed to find out their favorite seasons. The results are shown in the tally table. Use the tally table to respond to the following. How many more students chose winter than spring?

Our Favorite Seasons

Winter															
Spring							
Summer																
Fall				

Ⓐ 5 students
Ⓑ 10 students
Ⓒ 15 students
Ⓓ 17 students

19 The students in the third grade were surveyed to find out their favorite seasons. The results are shown in the tally table. Use the tally table to respond to the following. How many students were surveyed?

Our Favorite Seasons

Winter	ℍℍ ℍℍ ℍℍ II
Spring	ℍℍ II
Summer	ℍℍ ℍℍ ℍℍ III
Fall	III

Ⓐ 40 students
Ⓑ 45 students
Ⓒ 50 students
Ⓓ 55 students

20. Teachers at a nearby elementary school took a survey to determine how to plan for future field trips. What is the sum of the field trip choices that received the highest votes?

Field Trips	No. of Votes
Jersey Cape	11
turtle Back Zoo	7
camden's Children Garden	8
NJ Adventure Aquarium	11

Ⓐ 19
Ⓑ 18
Ⓒ 22
Ⓓ 15

21. The chart shows the range of snowfall expected each month at a local ski resort. During which month is there the greatest range in the amount of snowfall?

Month	Inches of Snow Fall
November	0 - 5
December	5 - 15
January	10 - 50
February	10 - 20

Ⓐ November
Ⓑ December
Ⓒ January
Ⓓ February

22. Based on the bar graph below, what can be said about the trend in bicycle helmet safety?

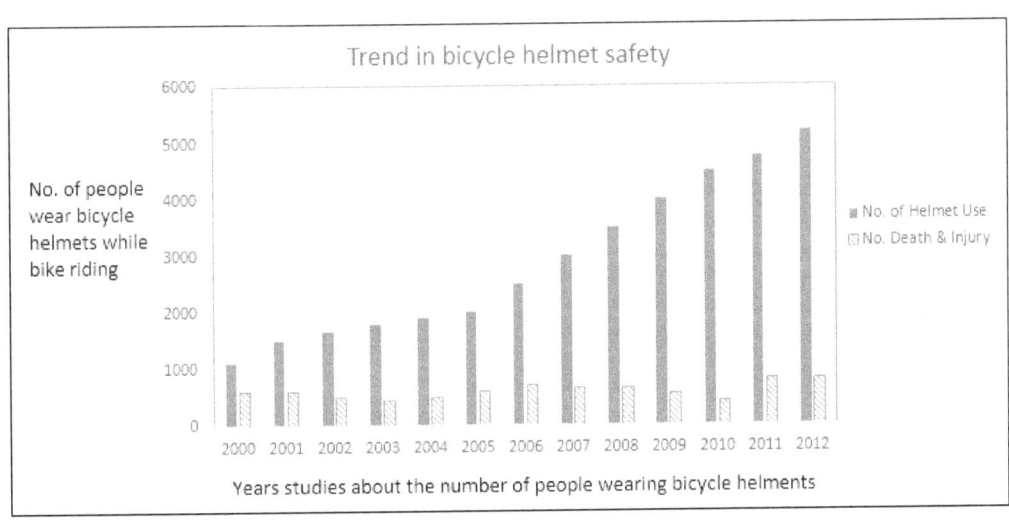

Ⓐ Bicycle riders who do not wear helmets are very unlikely to be injured or killed.
Ⓑ Bicycle riders who wear helmets are very unlikely to be injured or killed.
Ⓒ Wearing a helmet has no effect on bicycle safety.
Ⓓ Wearing a helmet affects bicycle safety on occasions.

23. The smoothie bar has several orders to fill. How many more orders call for sweet fruit smoothies than sour fruit smoothies?

Orders Call	Sweet Fruit Smoothies	Sour Fruit Smoothies
Ron Miller	10	2
Robert	5	5
George	4	3
Mike	6	4
Marisa	3	2
Greg	1	2

Ⓐ 23
Ⓑ 11
Ⓒ 12
Ⓓ 35

24. The line plot below shows the lengths of fish caught.

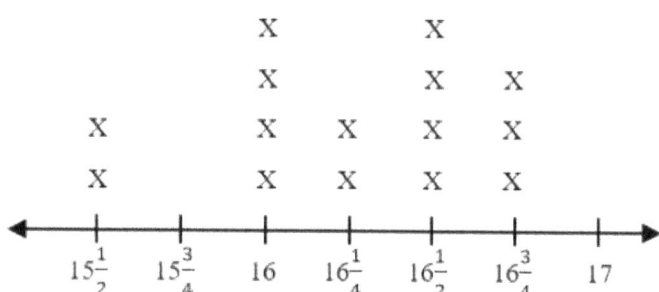

Lengths of Fish

Instruction: The line plot shows the data rounded to the nearest $\frac{1}{4}$ inch.

How much longer is the longest fish than the shortest fish? Drag and drop your answer into the box.

Ⓐ $1\frac{1}{2}$

Ⓑ $2\frac{1}{4}$

Ⓒ $1\frac{1}{4}$

Ⓓ $1\frac{3}{4}$

25. The line plot given below shows the lengths of fish caught. Find the total length if all the fish measuring $16\frac{1}{2}$ inches in length were to be laid end to end? Write your answer into the box.

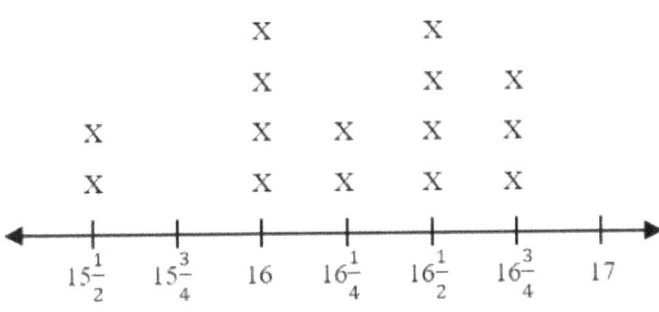

Lengths of Fish

Ⓐ 68 inches
Ⓑ 66 inches
Ⓒ 62 inches
Ⓓ 64 inches

26. Match each statement with the type of graph used for interpreting the data in each situation.

	Bar graph	Line graph	Pie Chart
The results of the number of boys in each grade.	○	○	○
The percentage of favorite desserts of the students in the class	○	○	○
The price of a car over the years.	○	○	○

27. Which group has the largest number of people in it? Write your answer in the box below

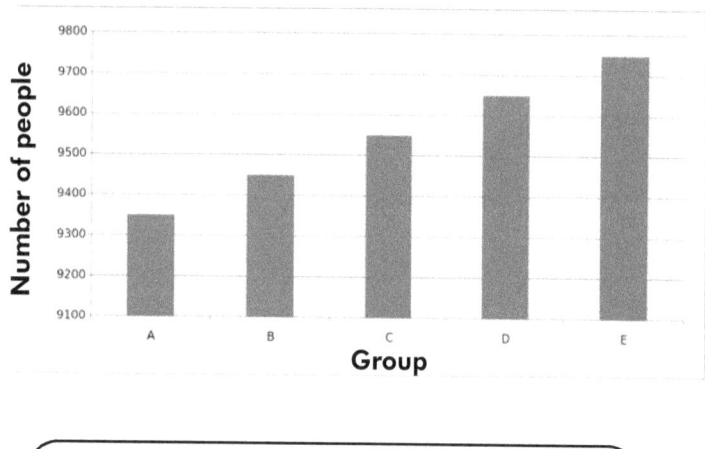

```
[                                                          ]
```

28. The line plot below shows the amount of sugar in 6 popular candy bars.

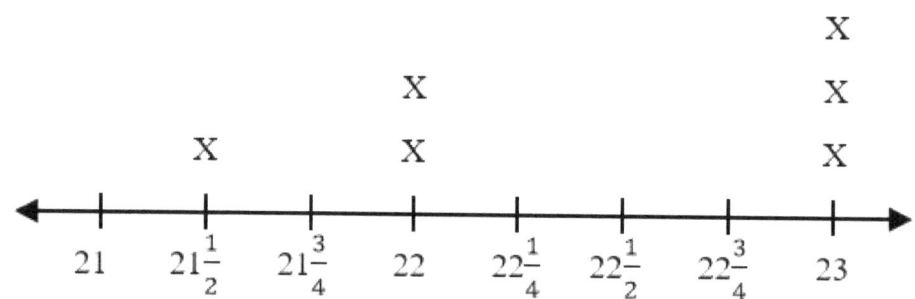

Instruction: The line plot shows the data rounded to the nearest $\frac{1}{2}$ g. How many total grams of sugar are in all 6 candy bars combined? Circle the correct answer.

Ⓐ $55\frac{1}{2}$ grams

Ⓑ $80\frac{1}{2}$ grams

Ⓒ $113\frac{1}{2}$ grams

Ⓓ $134\frac{1}{2}$ grams

29. Plot the number of people liking caramel apples, corn dogs and french fries in the form of a bar graph. Note: (each x represents one person.)

Favorite Carnival Foods

caramel apples	elephant ears	corn dogs	cotton candy	french fries	candy apples	funnel cakes
x			x			
x			x		x	x
x			x		x	x
x			x		x	x
x		x	x		x	x
x		x	x		x	x
x	x	x	x		x	x
x	x	x	x		x	x
x	x	x	x	x	x	x
x	x	x	x	x	x	x
x	x	x	x	x	x	x

30. Match the number of people liking the food to the correct category.
    Note: (each x represents one person.)

Favorite Carnival Foods

caramel apples	elephant ears	corn dogs	cotton candy	french fries	candy apples	funnel cakes
x			x			
x			x		x	x
x			x		x	x
x			x		x	x
x		x	x		x	x
x		x	x		x	x
x	x	x	x		x	x
x	x	x	x		x	x
x	x	x	x	x	x	x
x	x	x	x	x	x	x
x	x	x	x	x	x	x

Candy Apples -------------- (   )

Elephant Ears ------------- (   )

Cotton Candy -------------- (   )

Corn Dogs ---------------- (   )

CHAPTER 4 →Lesson 5: Angle Measurement

1. In the figure below, two lines intersect to form ∠A, ∠B, ∠C, and ∠D.
   If ∠C measures 128°, then ∠A measures:

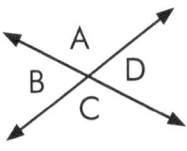

   Ⓐ 52°
   Ⓑ 128°
   Ⓒ 90°
   Ⓓ 180°

2. In the figure below, two lines intersect to form ∠A, ∠B, ∠C, and ∠D.
   If ∠C measures 128°, then ∠D measures:

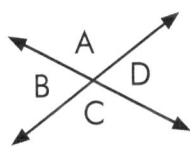

   Ⓐ 52°
   Ⓑ 128°
   Ⓒ 90°
   Ⓓ 180°

3. In the figure below, two lines intersect to form ∠A, ∠B, ∠C, and ∠D.
   If ∠B measures 68°, then ∠D measures:

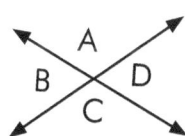

   Ⓐ 152°
   Ⓑ 90°
   Ⓒ 180°
   Ⓓ 68°

4. If the measurement of ∠A is 64°, then what is the measurement of ∠B?

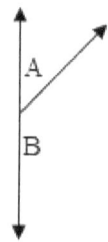

   Ⓐ 126°
   Ⓑ 119°
   Ⓒ 116°
   Ⓓ 64°

5. If the measurement of ∠B is 94°, then what is the measurement of ∠A?

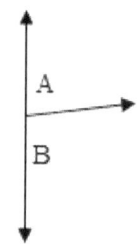

   Ⓐ 90°
   Ⓑ 86°
   Ⓒ 66°
   Ⓓ 94°

6 The total of the measures of ∠A and ∠B equals 90°. If ∠A measures 51°, then what does ∠B measure?

   Ⓐ 139°
   Ⓑ 51°
   Ⓒ 49°
   Ⓓ 39°

7. Two supplementary angles should total _____.

    Ⓐ 90°
    Ⓑ 45°
    Ⓒ 180°
    Ⓓ 360°

8. What should two complementary angles add up to?

    Ⓐ 45°
    Ⓑ 90°
    Ⓒ 180°
    Ⓓ 360°

9. Complete the sentence:
The measure of an obtuse angle is _____.

    Ⓐ less than the measure of a right angle
    Ⓑ equal to the measure of a right angle
    Ⓒ greater than the measure of a right angle
    Ⓓ less than the measure of an acute angle

10. What is the measure of the complement of an angle that measures 7°?

    Ⓐ 173°
    Ⓑ 87°
    Ⓒ 83°
    Ⓓ 73°

11. What is the measure of ∠A if ∠B measures 46° ?

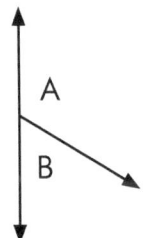

    Ⓐ 134°
    Ⓑ 143°
    Ⓒ 44°
    Ⓓ 90°

12. If ∠A measures 101° and ∠B measures 49° then what does ∠C measure?

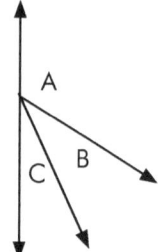

Ⓐ 150°
Ⓑ 20°
Ⓒ 40°
Ⓓ 30°

13. What is the supplement of an angle that measures 73°?

Ⓐ 117°
Ⓑ 107°
Ⓒ 23°
Ⓓ 17°

14. What is the measurement of each of the angles in this equilateral triangle?

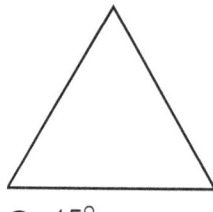

Ⓐ 45°
Ⓑ 60°
Ⓒ 90°
Ⓓ 180°

15. If ∠A measures 45° and ∠C measures 45° then what does ∠B measure?

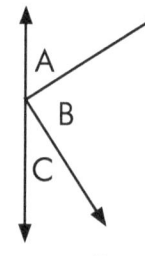

Ⓐ 45°
Ⓑ 75°
Ⓒ 90°
Ⓓ 135°

16. What is the angle measurement for the angle shown below? Write your answer in the box given below.

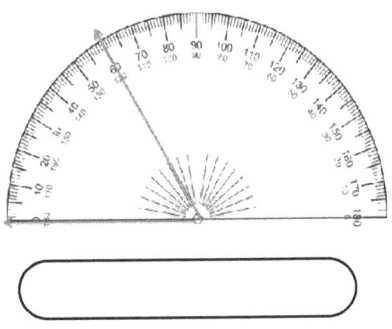

_____

17. What is the angle measurement for the below angle? Type the number in the box.

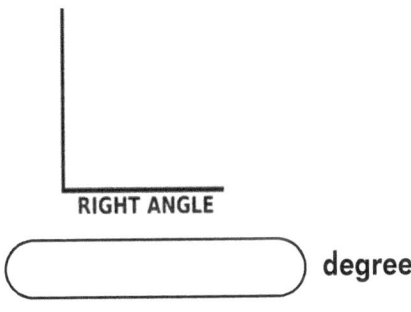

RIGHT ANGLE

(_____) degree

18. Which angles are obtuse angles? Select all the correct answers.

Ⓐ

Ⓑ

Ⓒ

Ⓓ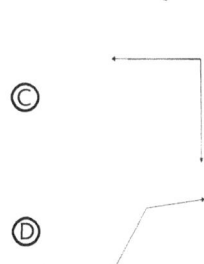

19. 1° is defined as 1/360 of a whole turn. What fraction of a whole turn is 45°?

Ⓐ 1/4
Ⓑ 1/8
Ⓒ 1/12
Ⓓ 1/6

CHAPTER 4 →Lesson 6: Measuring Turned Angles

1. At ice skating lessons, Erika attempts to do a 360 degree spin, but she only manages a half turn on her first attempt. How many degrees short of her goal was Erika's first attempt?

   Ⓐ 90 degrees
   Ⓑ 180 degrees
   Ⓒ 0 degrees
   Ⓓ 360 degrees

2. Erika's sister, Melanie, attempts to do a 360 degree turn, just like her sister, but she made a quarter turn on her first attempt. How many degrees short of her goal was Melanie's first attempt?

   Ⓐ 180 degrees
   Ⓑ 90 degrees
   Ⓒ 270 degrees
   Ⓓ 280 degrees

3. A water sprinkler covers 90 degrees of the Brown's backyard lawn. How many times will the sprinkler need to be moved in order to cover the full 360 degrees of the lawn?

   Ⓐ 4
   Ⓑ 2
   Ⓒ 3
   Ⓓ 5

4. A ceiling fan rotates 80 degrees and then it stops. How many more degrees does it need to rotate in order to make a full rotation?

   Ⓐ 265 degrees
   Ⓑ 90 degrees
   Ⓒ 180 degrees
   Ⓓ 280 degrees

5. Trixie is a professional photographer. She uses software on her computer to edit some wedding photos. She rotates the photograph of the bride 120 clockwise. Then she rotates it another 140 degrees clockwise. If she continues rotating the photo clockwise, how many more degrees will Trixie need to turn it to have made a complete 360 degree turn?

Ⓐ 100 degrees
Ⓑ 90 degrees
Ⓒ 180 degrees
Ⓓ 120 degrees

6. What is the angle measurement for the below angle? Type the number in the box.

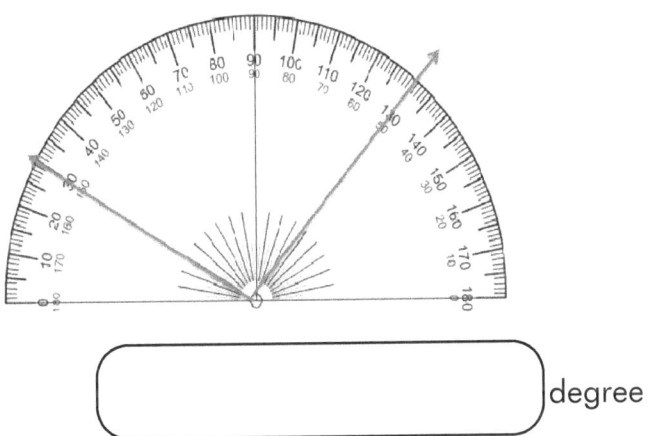

degree

7. Which angles are reflex angles? Select all the correct answers.

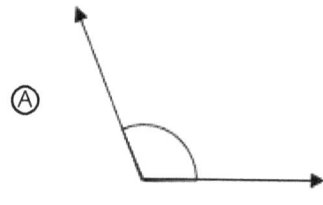

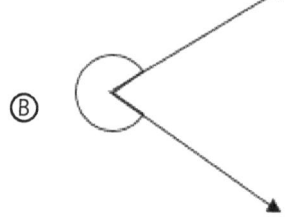

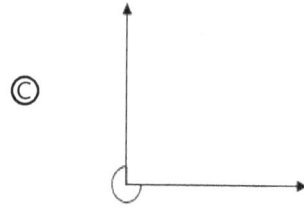

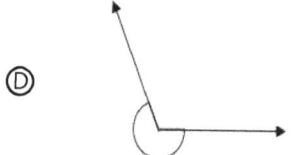

8. The sum of angle A and angle B is 260. If the measure of angle A is 120 degrees, what is the measure of angle B?

Ⓐ 60
Ⓑ 130
Ⓒ 140
Ⓓ 380

CHAPTER 4 →Lesson 7: Measuring and Sketching Angles

1. **What can be the measure of angle JKL?**

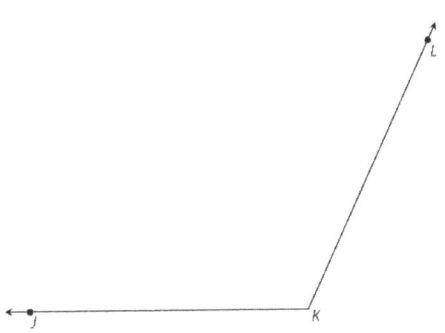

Ⓐ 65 degrees
Ⓑ 115 degrees
Ⓒ 75 degrees
Ⓓ 140 degrees

2. **What is the measure of this angle?**

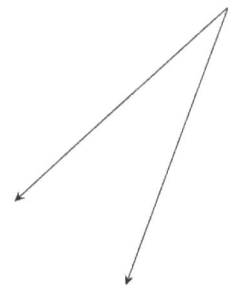

Ⓐ 10 degrees
Ⓑ 28 degrees
Ⓒ 48 degrees
Ⓓ 90 degrees

3. **What is the measure of angle PQR?**

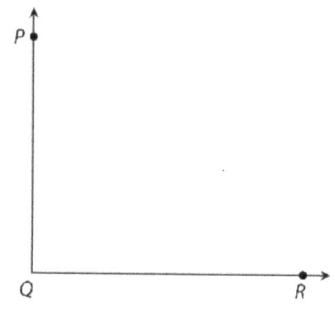

Ⓐ 360 degrees
Ⓑ 180 degrees
Ⓒ 0 degrees
Ⓓ 90 degrees

4. **What is the measure of the interior angles of this pentagon?**

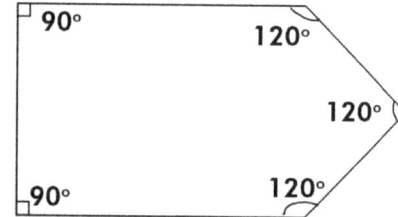

Ⓐ 400 degrees
Ⓑ 360 degrees
Ⓒ 540 degrees
Ⓓ 520 degrees

5. **What can you use to help measure angles?**

Ⓐ a degree
Ⓑ a protractor
Ⓒ a vertex
Ⓓ an angle

6. **What is the angle measurement for the below angle? Write the answer in the box below.**

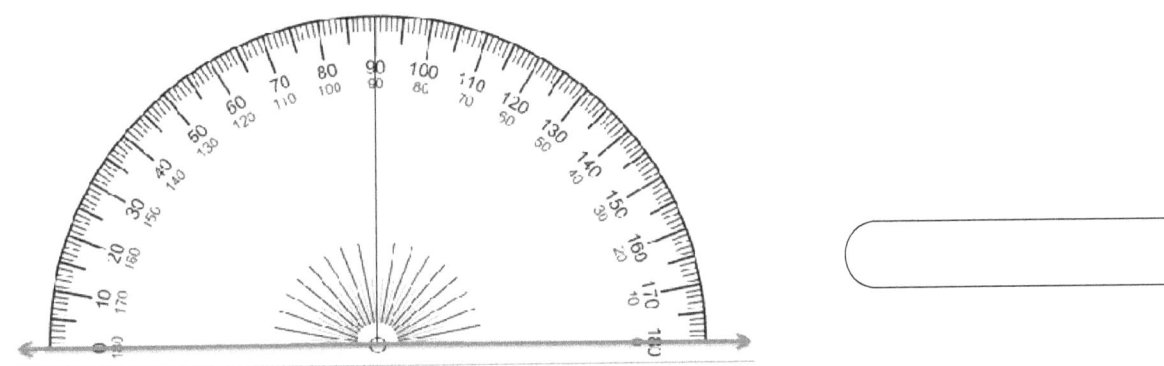

7. **What is the angle measurement for the below angle? Write the answer in the box below.**

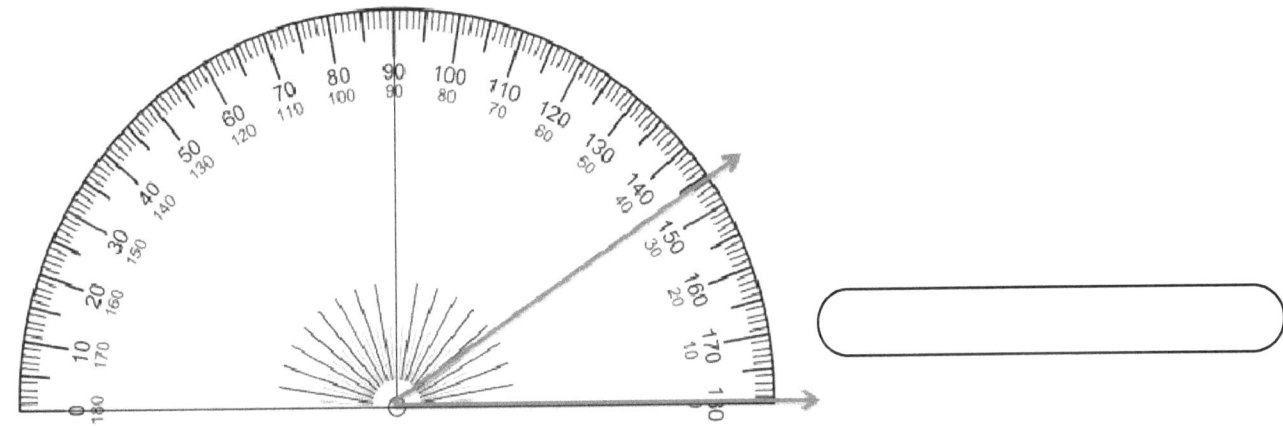

8. **What fraction of a whole turn is a straight angle? Circle the correct answer.**

Ⓐ 1/2
Ⓑ 1/4
Ⓒ 3/4
Ⓓ 5/8

9. **Which angle is equal to 1/6 of a whole turn? Shade the figure below to represent the answer.**

Instruction : 1 shaded cell = 10 degrees.

CHAPTER 4 →Lesson 8: Adding and Subtracting Angle Measurements

1. Angle 1 measures 40 degrees, and angle 2 measures 30 degrees. What is the measure of the angle PQR?

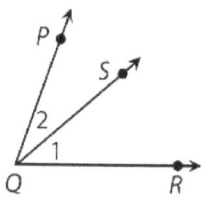

Ⓐ 70 degrees
Ⓑ 80 degrees
Ⓒ 100 degrees
Ⓓ 10 degrees

2. Angle ADC measures 120 degrees, and angle ADB measures 95 degrees. What is the measure of the angle BDC?

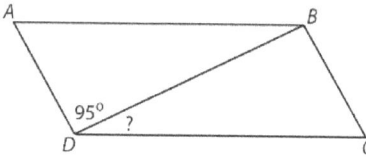

Ⓐ 5 degrees
Ⓑ 35 degrees
Ⓒ 15 degrees
Ⓓ 25 degrees

3. Angle JNM measures 100 degrees, angle JNK measures 25 degrees, and angle KNL measures 35 degrees. What is the measure of the angle LNM?

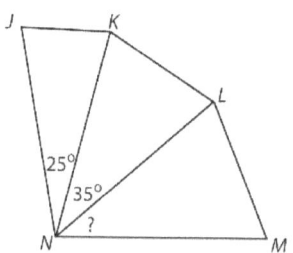

Ⓐ 35 degrees
Ⓑ 40 degrees
Ⓒ 30 degrees
Ⓓ 25 degrees

4. What fraction of a circle is a 180 degree angle?

　Ⓐ $\frac{1}{4}$

　Ⓑ $\frac{1}{2}$

　Ⓒ $\frac{1}{3}$

　Ⓓ $\frac{1}{5}$

5. What fraction of a circle is a 90 degree angle?

　Ⓐ $\frac{1}{4}$

　Ⓑ $\frac{1}{2}$

　Ⓒ $\frac{1}{3}$

　Ⓓ $\frac{1}{5}$

6. What is the angle measurement for the below angles when subtracted? Write the answer in the box shown below

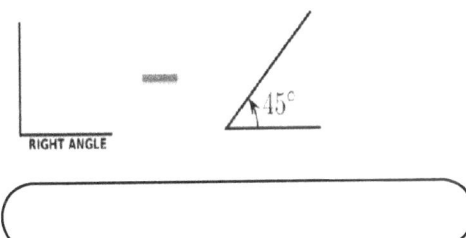

RIGHT ANGLE

7. What is the value of x° and y° in the figure below?

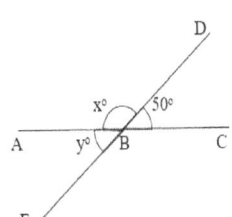

　Ⓐ x° = 120° and y° = 60°
　Ⓑ x° = 130° and y° = 50°
　Ⓒ x° = 50°　and y° = 130°
　Ⓓ x° = 110° and y° = 70°

End of Measurement & Data

Chapter 4
Measurement & Data

Lesson 1: Units of Measurement

Question No.	Answer	Detailed Explanation
1	A	Inches are not units of weight. Kilograms are not customary units. Tons are too large to use to measure a table. Pounds are the best choice.
2	B	Grams are not customary units. Between ounces and pounds, ounces are more reasonable unit to use.
3	A	12 lbs = 1.5 - 2 times a newborn baby. Gallons are used for liquid measurement. 200 ounces is only 12.5 pounds. 500 pounds is the most reasonable choice.
4	D	Meters and kilometers are not customary units. Yards are too small of a unit to measure the length of a road.
5	B	Gallons are not used to measure length. Centimeters are very small units. Miles are very large units. A tree's height would best be measured in yards.
6	A	Metric measurements are not customary units. Kilograms are part of the metric system.
7	B	Pounds measure weight, not liquid volume and yards measure length. Milliliters are part of the metric system, not the customary system of measurement.
8	C	There are 6 teaspoons in one fluid ounce. Multiply 6 x 2 to make a comparison.
9	B	There are 16 ounces in 1 pound. 128 oz = 8 pounds (too heavy) 200 oz = 12.5 pounds (too heavy) 2 ounces is way too light for a textbook.
10	B	There are 3 teaspoons in 1 tablespoon.
11	4	There are 1,000 meters in 1 kilometer so there are 4 kilometers in 4,000 meters.

Question No.	Answer	Detailed Explanation

12

	4	5	6
360 min = ___ hr			⬤
500 cm = ___ m		⬤	
240 sec = ___ min	⬤		

We know that there are 60 min in an hour, so we divide 360 by 60 to get the value for 360 min, which is 6 hours. We know there are 100 cm in 1 meter, so we divide 500 by 100 to get the value for 500 cm, which is 5 meters. We know there are 60 seconds in 1 min, so we divide 240 by 60 to get the value for 240 seconds, which is 4 minutes.

13

km	m	cm
6.5	6,500	650,000
8.2	**8,200**	**820,000**
7.3	7,300	**730,000**
8.25	**8,250**	825,000

1 km = 1,000 m, 1 m = 100 cm
6.5 km = 6.5 x 1,000 = 6,500 m
6,500 m = 6,500 x 100 = 650,000 cm
8.2 km = 8.2 x 1,000 = 8,200 m
8,200 m = 8,200 x 100 = 820,000 cm

$1 m = (\frac{1}{1,000})$ km ; $7,300 m = (\frac{7,300}{1,000}) = 7.3$ km

7,300 m = 7,300 x 100 = 730,000 cm

$1 cm = (\frac{1}{1,00})$ m ; $825,000 cm = (\frac{825,000}{100}) = 8,250$ m

$1 m = (\frac{1}{1,000})$ km ; $8,250 m = (\frac{8,250}{1,000}) = 8.250$ km = 8.25 km

14   A,D,E

1 pint = 2 cups; 44 pints = 44 x 2 cups = 88 cups
2 pints = 1 quart; 44 pints = 44 ÷ 2 quarts = 22 quarts
4 quarts = 1 gallon; 22 quarts = 22 ÷ 4 gallons = 5.5 gallons

Lesson 2: Measurement Problems

Question No.	Answer	Detailed Explanation
1	B	To determine the difference between the time Arthur wants to arrive at practice and how long it takes him to get there, subtract 45 minutes from 5:30 by counting backwards by 5's on the clock:
2	A	Correct answer is 9:43 PM. Add 2 hours to 7:05, which would make 9:05. Count forward by 5's on the face of a clock for 35 minutes, then add 3 more minutes to determine the exact time
3	B	There are 12 inches in 1 foot: Multiply 12 x 4, and then and add 5 more inches.
4	C	Add the prices of all the items and the amount of the taxes together. Subtract the sum from 300.00. $47.00 + 150.00 + 32.00 + 13.74 = $242.74 $300.00 - 242.74 = $57.26
5	C	One tablespoon equals $\frac{1}{16}$ of a cup. Therefore, to make $\frac{1}{2}$ cup you would need 8 tablespoons ($8 \times \frac{1}{16} = \frac{1}{2}$).
6	C	There are 16 ounces in 1 pound. Therefore, 32 ounces is more than 1 pound.
7	C	There are 12 months in 1 year. Calculate the total number of months and divide them by 12. The number of times 12 is divided by this number is the number of years, and the remainder is the number of months. 12 + 9 + 23 = 44 months 44 months = 3 years and 8 months
8	C	This is asking if the amount of flour used in the recipe is more or less than the amount of butter, and if the amount of butter is more or less than the amount of sugar. Check the given amounts to make a comparison.
9	B	The unit price is $3.25. Multiply that number by 4 to determine the cost of 4 books.
10	D	Multiply $24.99 x 3. Multiply $6.99 x 2. Add the products. $24.99 x 3 = $74.97 $6.99 x 2 = $13.98 $74.97 + 13.98 = $88.95

Question No.	Answer	Detailed Explanation
11	3	2 cups goes into 6 cups 3 times, so the answer is 3.

12

		$2\frac{3}{4}$	$4\frac{1}{4}$
	I need 5 cups of flour for my recipe. I need ¾ cup more. How much flour did I start with?		●
	I want to make a tug of war rope. It has to be 16 feet long. I have 18 feet 9 inches of rope. How much will I have to cut off? Express your answer as a mixed number in feet.	●	

I need 5 cups of flour for the recipe. I need $\frac{3}{4}$ cups more. So, I started with $5 - \frac{3}{4}$ = $4\frac{1}{4}$ cups of flour.

I have 18 feet 9 inches of rope. Tug of war rope's length = 16 feet. So, I have to cut off 18 feet 9 inches - 16 feet = 2 feet 9 inches. Since 12 inches = 1 feet, 9 inches = $\frac{9}{12}$ feet = $\frac{3}{4}$ feet. So, I have to cut off $2\frac{3}{4}$ feet.

Question No.	Answer	Detailed Explanation
13	67,17	To figure these out, you first need to figure out how much was done on each day. Then you need to add the three numbers together.

14   D

This is a problem on multiplication. Multiply the number of days (5) by the distance covered each day ($2\frac{1}{4}$ km).

$2\frac{1}{4} = \frac{(2 \times 4)+1}{4} = \frac{9}{4}$

Total distance covered $= 5 \times \frac{9}{4} = \frac{5 \times 9}{4} = \frac{45}{4}$ km

$1\,km = 1,000\,m;\ \frac{45}{4}\,km = \frac{45}{4} \times 1000 = \frac{45 \times 1000}{4}$

$= \frac{45000}{4} = \frac{(45000 \div 4)}{(4 \div 4)} = \frac{11250}{1} = 11,250\,m$

Lesson 3: Perimeter & Area

Question No.	Answer	Detailed Explanation
1	B	The formula for finding the area of a rectangle: Area = length x width.
2	A	The formula to find the perimeter of a rectangle: P = 2(length) + 2(width)
3	C	The length is 36 ft. and the the width is 48 ft.. All sides must be added to determine the perimeter. Add 48 + 36. Then multiply the sum by 2.
4	A	For the perimeter, add the measurements of the length and width: 72 + 30. Multiply the sum by 2.
5	A	Add the measurement of the length and the width, then multiply the sum by 2. Do this for both figures and compare the perimeters.
6	C	The perimeter is the sum of all 4 sides. If that sum is 100, divide 100 by 4 to determine the length of each side. (A square has 4 equal sides.)
7	D	To find the perimeter of a shape on the graph sheet, count each side of the squares all the way around the outside of the shape. The side of each square that makes up the shape is a unit.
8	B	To find the area of a shape on the graph sheet, count the number of squares on the inside of the shape.
9	C	Count the segments between the red dots, all the way around the shape.
10	B	Although all three of the choices offered would have an area of 48 square units, only a 12 by 4 rectangle would have a perimeter of 32 units.
11	20	To find the perimeter, add all the sides together. Since it is a rectangle, the 2 long sides are both 6, and the two short sides are 4. 6 + 6 + 4 + 4 = 20.

12

	100	56	102
Perimeter: ___ units		●	
Area: ___ units squared	●		

Perimeter is got by adding all the sides.

10+4+4+6+2+6+6+12+2+4= 56

The area is the sum of areas of triangles 1, 2, and 3, i.e.,
(10 x 4) + (12 x 4) + (6 x 2) = 40 + 48 + 12 = 100 sq units.

Question No.	Answer	Detailed Explanation
13	84,42	The area formula for a triangle is $\frac{1}{2}$ bh. The base is 14 and the height is 12, so we plug in the values into the equation: $\frac{1}{2}$ x 14 x 12 = 84. The perimeter formula is sum of each side. 13 + 14 + 15 = 42.
14	12 feet	A square is a special type of rectangle whose sides are of equal length. We know that the area of a rectangle = length x width. Therefore, the area of a square can be calculated by multiplying the length of one side by itself. Let A = area of a square, s = length of a side; A = s x s In this problem, A = 144; Therefore, we have 144 = s x s. What is the number when multiplied with itself gives 144? It is 12. Therefore, s = 12 feet

Lesson 4: Representing and Interpreting Data

Question No.	Answer	Detailed Explanation
1	D	The horizontal scale represents the number of cousins each student has. Read the number of x's plotted for 0 cousins.
2	C	Count the number of x's plotted above the number 4 on the horizontal scale.
3	B	The 2 foods that have the highest number of x's are the most favorites.
4	B	Subtract the number of families that pack before leaving for the picnic from the number of families that pack the night before the picnic. Each x represents 2 points. Therefore, the answer is (4 - 2) x 2 = 4.
5	A	The horizontal scale represents the days of the week. The vertical scale represents the number of office visits. The line on the graph started at 20 and continued to increase to almost 60.
6	B	The question is asking for a difference or comparison, which means subtraction. Find the number of visits for both days, then subtract: Monday = 20 from Friday = 55: 55 - 20.
7	B	Read the number that corresponds to the point plotted above Wednesday. If the point is halfway between numbers, add 5 points to the lower number the point sits between.
8	A	Add the total number of squares and triangles drawn on the graph. Each square represents 10 boxes and there are 14 squares shown: multiply 10 x 14. Each triangle represents 5 boxes and there are 3 triangles shown: 5 x 3. Calculate the products and add them together. Compare that sum to the total number of boxes needed to be sold.
9	B	To find the difference, total the number of boxes sold the first week, then the fourth week. Subtract the totals. Forty boxes were sold the first week; 55 boxes were sold the fourth week: 55 - 40.
10	D	To find the total, add the number of boxes sold both weeks. The first week, 40 boxes were sold; the second week, 35 boxes were sold: 40 + 35.
11	C	The horizontal scale represents the weeks that cans were sold. The vertical scale represents the number of cans sold. Compare the height of the bar to the number on the number scale for the second week. If a bar ends between two numbers, add 5 to the lower number the bar rests between.
12	B	Subtract to find the difference. Twenty-five cans were collected the second week; 40 were collected the third week: 40 - 25.

Question No.	Answer	Detailed Explanation
13	B	Compare the height of the bar to the number scale. Add 5 points to the lower number on the scale when the bar rests between numbers.
14	A	The week that has the shortest bar is the week with the fewest cans collected.
15	D	The weeks that show bars having the same height are the weeks that have the same number of collected cans.
16	C	The season showing the most tally marks is the most popular favorite season.
17	C	Count the number of tally marks shown for winter. Four tally marks with one tally crossing over them = 5 tally marks.
18	B	Subtract the number of tally marks shown for spring from the number of tally marks shown for winter.
19	B	To determine the total number of students who were surveyed or voted, add up the total number of tally marks shown on the graph.
20	C	Count the number of votes each field trip choice has. Add together the two field trip choices that have the most votes.
21	C	The range 10 - 50 for January is the greatest range shown on the chart.
22	B	The horizontal scale lists the years studies about the number of people wearing bicycle helmets while bike riding have been done. The vertical scale shows the number of people who wear bicycle helmets while bike riding. One bar shows the number of injuries and deaths that occurred while not wearing a helmet. The other bar shows the number of bike riders wearing helmets. Compare the height of the bars to the number scale to determine the number of people wearing bike helmets in each time frame and the number of injuries or deaths. Then compare the difference between these two sets of data.
23	B	Find the sum of all the sweet fruit smoothie orders, then find the sum of all the sour fruit smoothie orders. Subtract the number of sour fruit smoothie orders from the sweet fruit smoothie orders to determine the difference.

Question No.	Answer	Detailed Explanation
24	C	This is a problem on subtraction. Subtract the length of the shortest fish ($15\frac{1}{2}$ inches) from the length of the longest fish ($16\frac{3}{4}$ inches). Since the denominators of the fraction parts are different, write the equivalent fraction of $\frac{1}{2}$ with 4 as the denominator. $\frac{1}{2} = \frac{1\times2}{2\times2} = \frac{2}{4}$ $16\frac{3}{4} - 15\frac{1}{2} = 16\frac{3}{4} - 15\frac{2}{4} = 16 + \frac{3}{4} - (15 + \frac{2}{4})$ $= (16-15) + (\frac{3}{4} - \frac{2}{4}) = 1 + \frac{3-2}{4} = 1 + \frac{1}{4} = 1\frac{1}{4}$ The longest fish is longer by $1\frac{1}{4}$ inches than the shortest fish.
25	B	There are 4 fish which are $16\frac{1}{2}$ inches long. Therefore, total length when all the 4 fish are laid end to end. $= 4 \times 16\frac{1}{2}$ $= (4\times16) + 4\times\frac{1}{2}$ (using the distributive property) $= 64 + \frac{4}{2} = 64 + 2 = 66$ inches
26		

	Bar graph	Line graph	Pie Chart
The results of the number of boys in each grade.	◯		
The percentage of favorite desserts of the students in the class			◯
The price of a car over the years.		◯	

Bar graphs are best when looking at numbers in different categories, so that would be best used for counting the number of boys in each grade. Line graphs are best when looking at trends over a time span, so that would be best used for looking at the price of cars. Pie charts are best when looking at percentages, so that should be used when looking at the percentages of the favorite desserts.

Question No.	Answer	Detailed Explanation
27	E	The largest group will be the tallest bar, which is E.
28	D	There is one candy bar with sugar $21\frac{1}{2}$ g, 2 candy bars with sugar 22 g each and 3 candy bars with sugar 23 g each. Therefore, total amount of sugar $=21\frac{1}{2}+(2\times22)+(3\times23)$ $=21\frac{1}{2}+44+69$ $=21\frac{1}{2}+113$ $=(21+113)+\frac{1}{2}$ $=134+\frac{1}{2}$ $=134\frac{1}{2}$ g
29	11,7,3	Each "x" mark represents the number of people liking a particular food. So, for the 3 mentioned food categories, we get the correct value by counting the number of x marks. Then, the bar needs to be pulled up or down to represent the correct value.
30		The correct answers are : 1] Candy apples - 10 2] Elephant ears - 5 3] Cotton candy - 11 4] Corn dogs - 7

Lesson 5: Angle Measurement

Question No.	Answer	Detailed Explanation
1	B	Angle C and Angle A are congruent. They have the same exact measurements
2	A	Angle C and Angle D are supplementary. Their measures total 180°. 180 - 128 = 52
3	D	Angle B and Angle D are congruent so they will measure the same.
4	C	Together, Angle A and Angle B equal 180 degrees. Therefore, subtract 64 degrees from 180 to determine the measure of Angle B.
5	B	Subtract 94 from 180 to find the measure of Angle A.
6	D	To find the measure of Angle B, subtract 51 from 90.
7	C	Two supplementary angles together form a straight angle, which is equal to half of a circle and measures 180 degrees.
8	B	Two complementary angles total 90 degrees.
9	C	A right angle measures 90 degrees and looks like this: 90° This is an example of an obtuse angle: >90° <180°
10	C	Subtract 7 from 90. 90 - 7 = 83
11	A	Angle A and Angle B are supplementary. Subtract 46 from 180.
12	D	The measures of these three angles total 180 degrees. Add the measures of Angles A and B. Subtract the sum of those two angles from 180 to determine how much Angle C measures. 101 + 49 = 150 180 - 150 = 30
13	B	Subtract 73 from 180. 180 - 73 = 107
14	B	This is an equilateral triangle, in which all interior angles have the exact same measurement. Those angles total 180 degrees. 180 ÷ 3 = 60. Each angle measures 60 degrees.

Question No.	Answer	Detailed Explanation
15	C	All the angle measures must add up to 180 degrees. Add 45 and 45. Subtract the sum from 180. (Angle B is a right angle.)
16	60	To read the protractor, we look at how many degrees are in between the two sides. That value is 60°.
17	90	Since we know it is a right angle, we know it is 90 degrees
18	A&D	An obtuse angle measures more than 90 degrees but less than 180°. A and D are obtuse angles. B is an acute angle. C is a right angle (90°).
19	B	$45° = 45 \times \frac{1}{360}$ of a whole turn $45° = \frac{45 \times 1}{360}$ of a whole turn $= \frac{45}{360}$ of a whole turn $= \frac{45 \div 45}{360 \div 45}$ of a whole turn $= \frac{1}{8}$ of a whole turn. It means eight 45° angles make one whole turn (see the figure below)

Lesson 6: Measuring Turned Angles

Question No.	Answer	Detailed Explanation
1	B	A half-turn would be 180 degrees. Erika made a 180 degree turn, but her goal was 360 degrees. 360-180= 180 degrees.
2	C	Melanie's turn was 90 degrees. 360 (her goal) minus 90 (her turn) equals 270 degrees.
3	A	360 divided by 90 is 4. The sprinkler will need to be moved 4 times in order to cover the lawn.
4	D	360 degrees minus 80 degrees equals 280 degrees.
5	A	120 degrees plus 140 degrees equals 260 degrees. 360 degrees minus 260 degrees leaves 100 degrees left to make a full turn.
6	100	To read the protractor, we look at how many degrees are in between the two sides $130° - 30° = 100°$.
7	B,C,D	An obtuse angle measures more than 90° and measures less than 180°. A reflex angle measures more than 180° and measures less than 360°. A is an obtuse angle. B, C, and D are reflex angles
8	C	$\angle A + \angle B = 260$ $120 + \angle B = 260$. $\angle B = 260 - 120 = 140$ degrees.

Lesson 7: Measuring and Sketching Angles

Question No.	Answer	Detailed Explanation
1	B	Put the protractor with its center point on K, so that one ray points to 0°, the other ray points to 115°.
2	B	The angle is an acute angle and it is little less than 45 degrees (half of the right angle). Therefore, it cannot be 48 or 90 degrees. It is closer to 45 degrees. So, it cannot be 10 degrees. So, the reasonable answer is, the angle is 28 degrees.
3	D	Put the protractor with its center on the vertex and one ray pointing to 0°, so the other ray points to 90°.
4	C	The sum of the interior angles of a pentagon is 540°. This can be cross-checked by adding the angles 120 + 120 + 120 + 90 + 90 = 540.
5	B	Definition of protractor: a tool used to measure angles.
6	180	To read the protractor, we look at how many degrees are in between the two sides. That value is 180°.
7	35	To read the protractor, we look at how many degrees are in between the two sides. That value is 35°. Be careful because you have to read the blue numbers since the angle starts at the blue 0.
8	A	A straight angle measures 180°. $1o$ is defined as $\dfrac{1}{360°}$ of a whole turn. Therefore, $180° = 180 \times \dfrac{1}{360°}$ of a whole turn $180° = \dfrac{180 \times 1}{360°}$ of a whole turn $= \dfrac{180}{360}$ of a whole turn $= \dfrac{(180 \div 180)}{(360 \div 180)}$ of a whole turn $= \dfrac{1}{2}$ of a whole turn. It means two 180° angles make one whole turn.

Question No.	Answer	Detailed Explanation
9		Let the measure of the angle which makes $\frac{1}{6}$ of a whole turn be S. We know that one whole turn is equal to 360°. Therefore, the measure of the angle which makes $\frac{1}{6}$ of a whole turn can be calculated by multiplying $\frac{1}{6}$ by 360°. Measure of angles $S = \frac{1}{6} \times 360° = \frac{1 \times 360°}{6} = \frac{360°}{6} = 60°$. One shaded cell = 10 degrees. Therefore, we have to shade 60 ÷ 10 = 6 cells.

Lesson 8: Adding and Subtracting Angle Measurements

Question No.	Answer	Detailed Explanation
1	A	The measure of the angle PQR is the sum of angle 1 & angle 2, so it is 40 + 30 = 70.
2	D	The measure of angle BDC can be found by subtracting 95 degrees from 120 degrees, for 25 degrees
3	B	Measure of Angle LNM = 100 - (25 + 35) = 100 - 60 = 40
4	B	1/2 of a circle is 180 degrees.
5	A	1/4 of a circle is 90 degrees.
6	45	We know a right angle is 90°, and it tells us the next angle is 45, so we know 90° - 45° = 45°
7	B	The sum of the measures of the angles ∠DBC and ∠ABD is equal to the measure of the ∠ABC. ∠ABC is a straight angle. It measures 180°. Therefore, ∠ABD + ∠DBC = 180° x° = 180° - ∠DBC = 180° - 50° = 130° The sum of the measures of the angles ∠ABE and ∠ABD is equal to the measure of the ∠EBD. ∠EBD is a straight angle. It measures 180°. Therefore, ∠ABE + ∠ABD = 180° y° = 180° - ∠ABD = 180° - 130° = 50°

Chapter 5: Geometry

Lesson 1: Points, Lines, Rays, and Segments

1. **Which of the following is a quadrilateral?**

   Ⓐ Triangle
   Ⓑ Rhombus
   Ⓒ Pentagon
   Ⓓ Hexagon

2. **How many sides does a pentagon have?**

   Ⓐ 3
   Ⓑ 2
   Ⓒ 1
   Ⓓ 5

3. **Use the network below to respond to the following question: How many line segments connect directly to Vertex F?**

   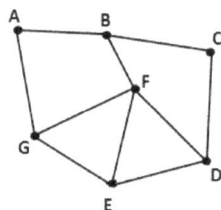

   Ⓐ 3
   Ⓑ 4
   Ⓒ 5
   Ⓓ 6

4. **Which of these statements is true?**

   Ⓐ A parallelogram must be a rectangle.
   Ⓑ A trapezoid might be a square.
   Ⓒ A rhombus must be a trapezoid.
   Ⓓ A rectangle must be a parallelogram.

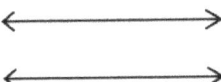

5. **What is being shown below?**

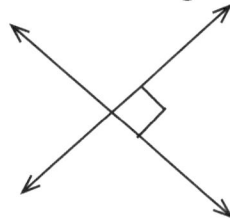

Ⓐ a pair of parallel lines
Ⓑ a pair of intersecting lines
Ⓒ a pair of congruent rays
Ⓓ a pair of perpendicular lines

6. **What is being shown below?**

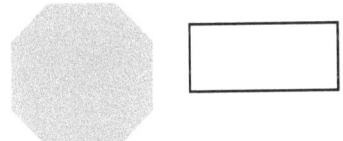

Ⓐ a pair of parallel lines
Ⓑ a pair of perpendicular lines
Ⓒ a pair of obtuse angles
Ⓓ a net for a cube

7. **Identify the following plane figures.**

Ⓐ heptagon and quadrilateral
Ⓑ hexagon and quadrilateral
Ⓒ octagon and quadrilateral
Ⓓ octagon and pentagon

8. **How many sides does an octagon have?**

Ⓐ 8
Ⓑ 7
Ⓒ 6
Ⓓ 5

9. **What kind of lines intersect to make 90 degree angles?**

Ⓐ supplementary
Ⓑ perpendicular
Ⓒ skew
Ⓓ parallel

10. **How many right angles does a parallelogram have?**

Ⓐ 0
Ⓑ 2
Ⓒ 4
Ⓓ It depends on the type of parallelogram.

11. **Circle the 'Point' from the figures.**

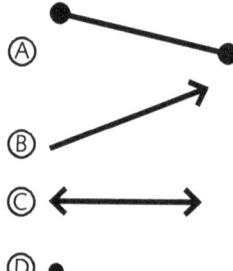

12. **How many end points does a segment have? Write your answer in the box below.**

13. **With reference to the figure below, which of the following statements are correct? Select all the correct answers.**

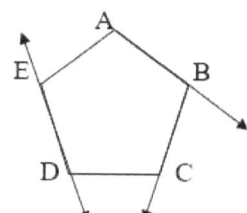

Ⓐ There are 4 rays.
Ⓑ There are 3 line segments.
Ⓒ There is one line.
Ⓓ There is no line.

CHAPTER 5 →Lesson 2: Angles

1. The hands of this clock form a/an _____ angle.

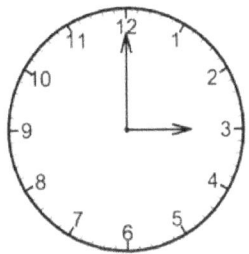

Ⓐ obtuse
Ⓑ straight
Ⓒ right
Ⓓ acute

2. How many acute angles and obtuse angles are there in the figure shown below?

Ⓐ 2 acute angles and 6 obtuse angles
Ⓑ 4 acute angles and 4 obtuse angles
Ⓒ 8 acute angles and 0 obtuse angles
Ⓓ 0 acute angles and 8 obtuse angles

3. Describe the angles found in this figure.

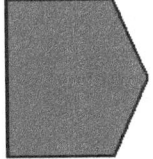

Ⓐ 2 right angles and 3 acute angles
Ⓑ 3 right angles and 2 obtuse angles
Ⓒ 2 right angles, 2 obtuse angles, and 1 acute angle
Ⓓ 2 right angles and 3 obtuse angles

4. **What type of triangle is shown below?**

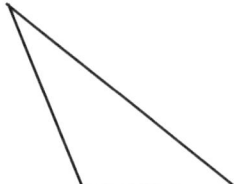

Ⓐ Isosceles triangle
Ⓑ Scalene triangle
Ⓒ Equilateral triangle
Ⓓ None of the above

5. **Which statement is true about an obtuse angle?**

Ⓐ It measures less than 90 degrees.
Ⓑ It measures more than 90 degrees.
Ⓒ It measures exactly 90 degrees.
Ⓓ It measures more than 180 degrees.

6. **Which statement is true about a straight angle?**

Ⓐ It measures less than 90 degrees.
Ⓑ It measures more than 90 degrees.
Ⓒ It measures exactly 90 degrees.
Ⓓ It measures exactly 180 degrees.

7. **A square has what type of angles?**

Ⓐ 2 acute and 2 right angles
Ⓑ 4 right angles
Ⓒ 4 acute angles
Ⓓ It depends on the size of the square.

8. **Which statement is true about an equilateral triangle.**

Ⓐ It has 1 acute angle and 2 obtuse angles.
Ⓑ It has 2 acute angles and 1 right angle.
Ⓒ It has all 30 degree angles.
Ⓓ It has all 60 degree angles.

9. **Classify the angle:**

Ⓐ right
Ⓑ acute
Ⓒ obtuse
Ⓓ straight

10. Tina wanted her bedroom area rug to be designed after a geometric shape. The rug has somewhat of a circular shape with 7 straight sides and 7 obtuse angles. What is the name of this shape?

Ⓐ hexagon
Ⓑ octagon
Ⓒ pentagon
Ⓓ heptagon

11. **Select all the obtuse angles.**

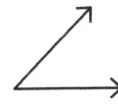   

12. How many of the following angles are right angles? Write your answer in the box below.

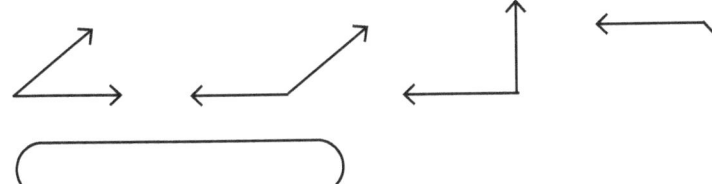

13. Choose the letters among the following which have at least one right angle. Note that there may be more than one correct answer.

Ⓐ L
Ⓑ T
Ⓒ V
Ⓓ E

CHAPTER 5 →Lesson 3: Classifying Plane (2-D) Shapes

1. **Complete the sentence:**
   A polygon is named based on _____.

   Ⓐ how many sides or interior angles it has
   Ⓑ how many of its sides are straight
   Ⓒ how large it is
   Ⓓ how many lines of symmetry it has

2. **Complete the sentence:**
   A polygon with 4 sides and 4 interior angles is called a/an _____.

   Ⓐ triangle
   Ⓑ quadrilateral
   Ⓒ pentagon
   Ⓓ octagon

3. **Complete the sentence:**
   A rectangle must have _____ .

   Ⓐ all parallel sides and all congruent sides
   Ⓑ 2 pairs of parallel sides and 2 pairs of congruent sides
   Ⓒ 2 pairs of parallel sides and 4 congruent sides
   Ⓓ 1 pair of parallel sides and 1 pair of congruent sides

4. **Complete the sentence:**
   A polygon must have _____ .

   Ⓐ 3 or more interior angles
   Ⓑ at least one pair of parallel sides
   Ⓒ at least one pair of congruent sides
   Ⓓ a line of symmetry

5. Classify this triangle:

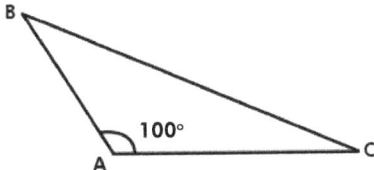

   Ⓐ acute triangle
   Ⓑ obtuse triangle
   Ⓒ right triangle
   Ⓓ straight triangle

6. If the angles on a triangle all measure less than 90 degrees, what type of triangle is it?

   Ⓐ obtuse triangle
   Ⓑ right triangle
   Ⓒ straight triangle
   Ⓓ acute triangle

7. Another name for a 180 degree angle is a/an _____.

   Ⓐ right angle
   Ⓑ obtuse angle
   Ⓒ straight angle
   Ⓓ acute angle

8. Complete the sentence:
A point has _____.

   Ⓐ size but no position
   Ⓑ size and position
   Ⓒ neither size nor position
   Ⓓ position but no size

9. **Name the angle found at the corners of the swimming pool shown below.**

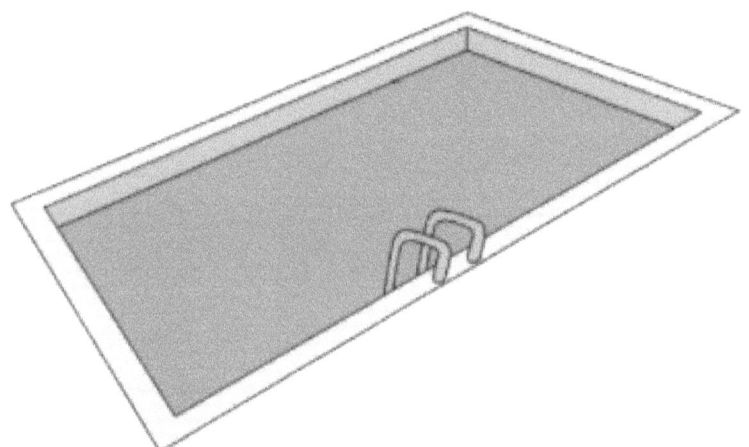

Ⓐ Angles found at the corners of the swimming pool consist of acute angles.
Ⓑ Angles found at the corners of the swimming pool consist of straight angles.
Ⓒ Angles found at the corners of the swimming pool consist of right angles.
Ⓓ Angles found at the corners of the swimming pool consist of obtuse angles.

10. **Classify the triangles shown in this design.**

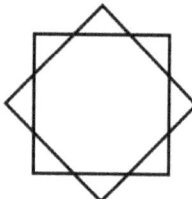

Ⓐ They are acute triangles.
Ⓑ They are right triangles.
Ⓒ They are obtuse triangles.
Ⓓ They are straight triangles.

11. **Select all the pentagons by circling them.**

12. How many triangles can be found in this figure? Write your answer in the box below.

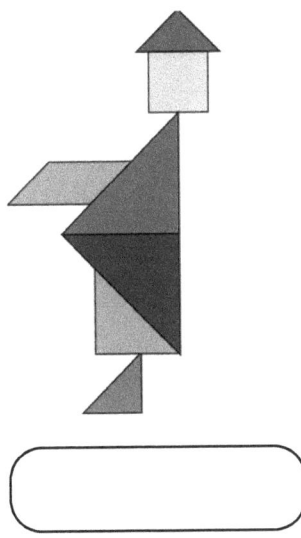

13. In the first column, classification of the triangles based on the lengths of the sides is given. Select all possible triangles for each type of triangle. (if the answer is "possible" select that option by ticking the option)

	can be an acute triangle	can be an obtuse triangle	can be a right triangle
An equilateral triangle	☐	☐	☐
An isosceles triangle	☐	☐	☐
A scalene triangle	☐	☐	☐

14. **For each statement given in the first column, select True if the statement is correct, or select False if it is incorrect**

	True	False
An acute triangle can be an equilateral triangle	○	○
An acute triangle cannot be an isosceles triangle	○	○
An acute triangle cannot be a scalene triangle	○	○
All right triangles are scalene triangles	○	○
An obtuse triangle can be an isosceles triangle	○	○
An obtuse triangle can be a scalene triangle	○	○

CHAPTER 5 →Lesson 4: Symmetry

1. How many lines of symmetry does an equilateral triangle have?

   Ⓐ 1
   Ⓑ 3
   Ⓒ 2
   Ⓓ 0

2. How many lines of symmetry does a regular pentagon have?

   Ⓐ 1
   Ⓑ 2
   Ⓒ 5
   Ⓓ 10

3. How many lines of symmetry does a rectangle have?

   Ⓐ 1
   Ⓑ 2
   Ⓒ 3
   Ⓓ 4

4. How many lines of symmetry does a regular octagon have?

   Ⓐ 2
   Ⓑ 4
   Ⓒ 6
   Ⓓ 8

5. How many lines of symmetry does a square have?

   Ⓐ 2
   Ⓑ 4
   Ⓒ 6
   Ⓓ 8

6. How many lines of symmetry does a parallelogram have if it is not a square, rectangle, or rhombus?

   Ⓐ 2
   Ⓑ 4
   Ⓒ 1
   Ⓓ 0

7. How many lines of symmetry does the following object have?

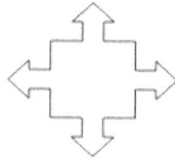

- Ⓐ 0
- Ⓑ 2
- Ⓒ 4
- Ⓓ 6

8. How many lines of symmetry does the following object have?

- Ⓐ 0
- Ⓑ 1
- Ⓒ 2
- Ⓓ 3

9. How many lines of symmetry does the following shape have?

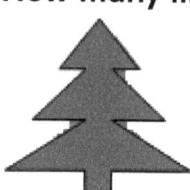

- Ⓐ 4
- Ⓑ 3
- Ⓒ 2
- Ⓓ 1

10. Regular quadrilaterals have 4 angles and 4 lines of symmetry. Regular pentagons have 5 angles and 5 lines of symmetry. Regular hexagons have 6 angles and 6 lines of symmetry. If this pattern were to continue, how many lines of symmetry does a regular heptagon have?

- Ⓐ 7
- Ⓑ 8
- Ⓒ 3
- Ⓓ 2

11. Select the heart with the correct line of symmetry below by circling it.

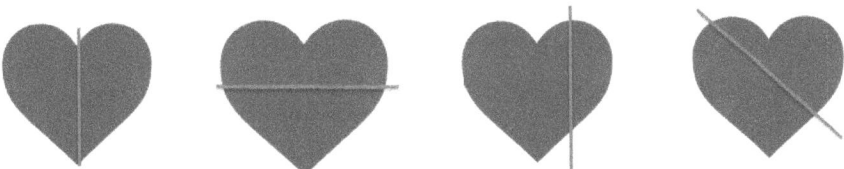

12. Select the figures with the correct lines of symmetry. Note that more than one option may be correct. Select all the correct answers.

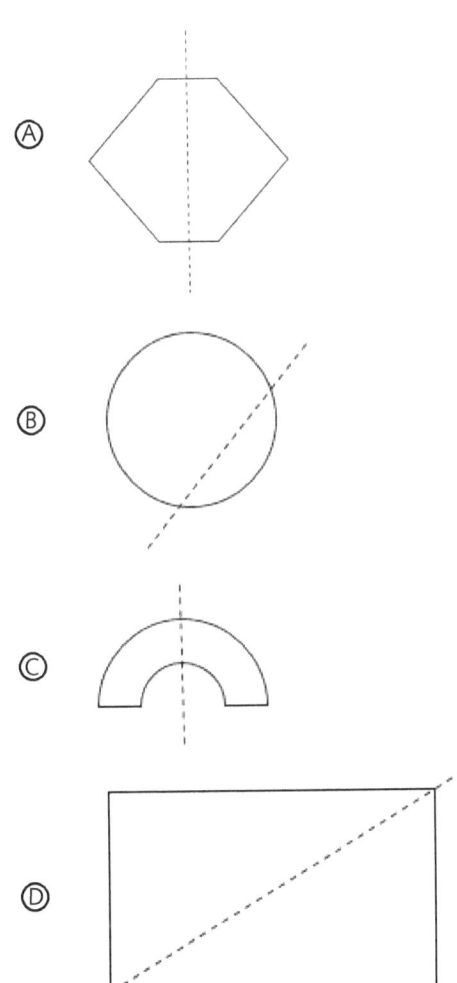

End of Geometry

Chapter 5: Geometry

Lesson 1: Points, Lines, Rays, and Segments

Question No.	Answer	Detailed Explanation
1	B	Quadrilaterals are 4 sided-polygons. The prefix "quad" means 4. A rhombus has 4 sides.
2	D	The prefix "pent" means 5.
3	B	Vertices are points where two line segments meet, or intersect. Four line segments meet at Point F.
4	D	Vertices are points where two line segments meet, or intersect. Four line segments meet at Point F.
5	A	Lines are drawn with arrows on each end. If two lines cross at right angles, they are perpendicular. If two lines never cross, they are called parallel.
6	B	These two lines cross to form right angles. By definition, the lines are perpendicular.
7	C	A polygon is named after the number of sides and interior angles it has: "tri" means 3, "quad" means 4, "pent" means 5, "hex" means 6, "hept" means 7 and "oct" means 8.
8	A	The prefix in front of the word octagon is "oct" and means 8.
9	B	By definition, perpendicular lines must cross to form right (90-degree) angles. Perpendicular 90°
10	D	A parallelogram is a Quadrilaterals with two pairs of parallel sides. There is no requirement for it to have a certain number of right angles. It may have 0 or it may have 4.

Question No.	Answer	Detailed Explanation
11	D	● A point is defined as a basic figure in geometry that has a location, but no size. Among the choices given, last one is the point.
12	2	A segment is a portion of a line, but has two ends, so there are 2 end-points
13	A & C	A ray is straight and has one endpoint. In the figure, AB, BC, DE, and ED are rays. Therefore, option (A) is correct. A line segment is straight and has two endpoints. In the figure, AE and DC are line segments. Therefore, option (B) is incorrect. A line is straight and extends infinitely in both directions. In the figure, DE is a line. Therefore, option (C) is correct. DE is a line. Therefore, (D) is incorrect.

Lesson 2: Angles

Question No.	Answer	Detailed Explanation
1	C	The hands of this clock differ by 90 degrees. So, the answer is Right angle.
2	D	Obtuse angles are larger than 90 degrees; acute angles are smaller than 90 degrees.
3	D	Right angles are 90 degrees, acute angles are less than 90 degrees and obtuse angles are more than 90 degrees.
4	B	This triangle has no equal sides. An equilateral triangle has all equal sides; an isosceles triangle has 2 equal sides.
5	B	An obtuse angle measures between 90 and 180 degrees.
6	D	A straight angle measures the same as 2 right angles. It is half of a full circle, which is 360 degrees.
7	B	Each angle of a square measures 90 degrees, regardless of its size.
8	D	An equilateral triangle is a closed shape that has all equal angles that add up to 180 degrees. This means each of the three angles measures 60°.
9	B	This angle measures less than 90 degrees.
10	D	The prefix "hept" means 7. A heptagon is a polygon with 7 sides and 7 angles.
11		Obtuse angles are angles that are greater than 90 degrees. The last three angles are all greater than 90 degrees.
12	1	A right angle is 90 degrees, or a corner to a square. The third angle is the only one that is a right angle.
13	A,B & D	The letter L has one right angle, the letter T has two right angles, and the letter E has three right angles.

Lesson 3: Classifying Plane (2-D) Shapes

Question No.	Answer	Detailed Explanation
1	A	The names given to different types of polygons (triangle, quadrilateral, pentagon, etc.) are based on how many sides or interior angles each figure has.
2	B	The prefix "quad" means 4.
3	B	The opposite sides of a rectangle must be congruent and parallel. In addition to this, each of the 4 angles of a rectangle measure 90 degrees.
4	A	Triangles, quadrilaterals, pentagons, hexagons, heptagons and octagons are examples of polygons. They all have at least 3 interior angles.
5	B	This triangle has an angle that measures more than 90 degrees. Therefore, it is classified as an obtuse triangle.
6	D	An acute triangle has three angles that each measure less than 90 degrees.
7	C	A straight angle is an angle that measures exactly 180 degrees.
8	D	A point is just a location. It has a position in space, but no size.
9	C	Many of the angles in this photo measure 90 degrees. There are mostly right angles (the corners of the house, the door and window frames, the steps in the pool).
10	B	The triangles in this design have right angles. Therefore, they would be classified as right triangles.
11		Pentagons are shapes with 5 sides. There are 2 pentagons above: the first shape and the third shape.
12	6	Triangles are shapes with 3 sides. There are 6 triangles above.

Question No.	Answer	Detailed Explanation
13		

	can be an acute triangle	can be an obtuse triangle	can be a right triangle
An equilateral triangle	○		
An isosceles triangle	○	○	○
A scalene triangle	○	○	○

(a) An equilateral triangle is always an acute triangle. Because, each angle in an equilateral triangle measures 60°.

(b) An isosceles triangle can be any of the three triangles: acute triangle, obtuse triangle or right triangle. Examples : A triangle whose angles measure 70°, 70° and 40° is an isosceles acute triangle. A triangle whose angles measure 110°, 35° and 35° is an isosceles obtuse triangle. A triangle whose angles measure 90°, 45° and 45° is an isosceles right triangle.

(c) A scalene triangle can be any of the three triangles: acute triangle, obtuse triangle or right triangle. Examples : A triangle whose angles measure 70°, 60° and 50° is a scalene acute triangle. A triangle whose angles measure 110°, 40° and 30° is a scalene obtuse triangle. A triangle whose angles measure 90°, 60° and 30° is a scalene right triangle.

Question No.	Answer	Detailed Explanation		
14			true	false
		An acute triangle can be an equilateral triangle	○	
		An acute triangle cannot be an isosceles triangle		○
		An acute triangle cannot be a scalene triangle		○
		All right triangles are scalene triangles		○
		An obtuse triangle can be an isosceles triangle	○	
		An obtuse triangle can be a scalene triangle	○	

An acute triangle can be any of the three triangles: equilateral or isosceles or scalene. Examples : An acute triangle whose angles measure 60°, 60° and 60° is an equilateral triangle. An acute triangle whose angles measure 80°, 80° and 20° is an acute isosceles triangle. An acute triangle whose angles measure 75°, 65° and 40° is an acute scalene triangle.

A right triangle can be an isosceles triangle or a scalene triangle (It cannot be an equilateral triangle because each angle in an equilateral triangle measures 60°). A right triangle whose angles measure 90°, 45° and 45° is a right isosceles triangle. A right triangle whose angles measure 90°, 65° and 25° is a right scalene triangle.

An obtuse triangle can be an isosceles triangle or a scalene triangle (It cannot be an equilateral triangle because each angle in an equilateral triangle measures 60°). An obtuse triangle whose angles measure 120°, 30° and 30° is an obtuse isosceles triangle. An obtuse triangle whose angles measure 130°, 30° and 20° is an obtuse scalene triangle.

Lesson 4: Symmetry

Question No.	Answer	Detailed Explanation
1	B	A line of symmetry is an imaginary line that separates a figure into identical parts.
2	C	For regular polygons, the number of lines of symmetry equals the number of sides the shape has. A pentagon has 5 sides, so there are 5 lines of symmetry.
3	B	A rectangle has two lines of symmetry. One is horizontal through its center, the other is vertical through its center.
4	D	An octagon is an 8 sided polygon. Therefore, a regular octagon would have 8 lines of symmetry.
5	B	A square is actually a regular quadrilateral. Therefore, it must have four lines of symmetry.
6	D	Irregular parallelograms are slanted. Therefore, the two sides will not match up when folded. There are no lines of symmetry.
7	C	A line of symmetry divides a figure into identical parts or mirror images of each other. This figure has one horizontal, one vertical, and two diagonal lines of symmetry.
8	B	One line can be drawn vertically through the center of this shape to divide it into two matching parts.
9	D	The point at the top is perfectly centered. One line can be drawn vertically through this point to divide the shape into two matching parts.
10	A	Each mentioned polygon increases by one angle and line of symmetry. The prefix "hept" means 7.
11		The line of symmetry is the line that is where you can fold and have two identical halves.
12	A ,C & D	If a figure can be folded along the line into matching parts, then it is a line of symmetry. The figures (A) and (C) can be folded along the dashed lines into matching parts. The figures (B) and (D) cannot be folded along the dashed lines into matching parts.

Notes

Test Taking Tips

1) **The day before the test,** make sure you get a good night's sleep.

2) **On the day of the test,** be sure to eat a good hearty breakfast! Also, be sure to arrive at school on time.

3) **During the test:**

- **Read every question carefully.**

  - Do not spend too much time on any one question. Work steadily through all questions in the section.
  - Attempt all of the questions even if you are not sure of some answers.
  - If you run into a difficult question, eliminate as many choices as you can and then pick the best one from the remaining choices. Intelligent guessing will help you increase your score.
  - Also, mark the question so that if you have extra time, you can return to it after you reach the end of the section.
  - Some questions may refer to a graph, chart, or other kind of picture. Carefully review the graphic before answering the question.
  - Be sure to include explanations for your written responses and show all work.

- **While Answering Multiple-Choice (EBSR) questions.**

  - Select the bubble corresponding to your answer choice.
  - Read **all** of the answer choices, even if think you have found the correct answer.

- **While Answering TECR questions.**

  - Read the directions of each question. Some might ask you to drag something, others to select, and still others to highlight. Follow all instructions of the question (or questions if it is in multiple parts)

Frequently Asked Questions(FAQs)

For more information on the Assessment, visit
www.lumoslearning.com/a/wytopp-faqs
OR Scan the **QR Code**

Step 1 → **Visit the link given below and login to your parent/teacher account**
www.lumoslearning.com

Step 2 → Open the **My Account** menu, go to **My Subscriptions**, and select **My tedBooks**. Enter the book access code shown on the first page of the book and submit.

Step 3 → **Add the new book**

To add the new book for a registered student, choose the '**Student**' button and click on submit.

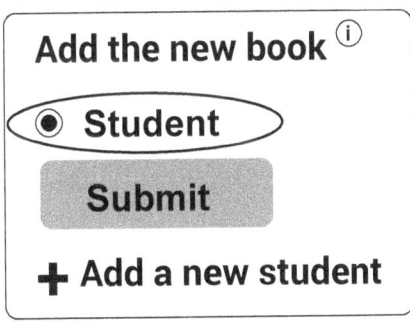

To add the new book for a new student, choose the '**Add New Student**' button and complete the student registration.

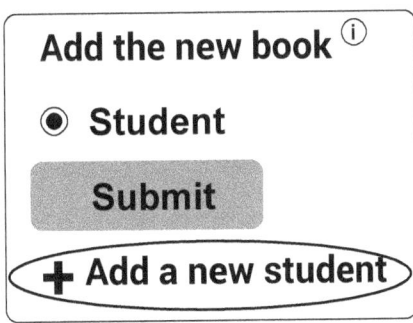

Progress Chart

Standard	Lesson	Score	Date of Completion
4.OA.A.1	Number Sentences		
4.OA.A.2	Real World Problems		
4.OA.A.3	Multi-Step Problems		
4.OA.B.4	Number Theory		
4.OA.C.5	Patterns		
4.NBT.D.1	Place Value		
4.NBT.D.2	Compare Numbers and Expanded Notation		
4.NBT.D.3	Rounding Numbers		
4.NBT.E.4	Addition & Subtraction		
4.NBT.E.5	Multiplication		
4.NBT.E.6	Division		
4.NF.F.1	Equivalent Fractions		
4.NF.F.2	Compare Fractions		
4.NF.G.3.A	Adding & Subtracting Fractions		
4.NF.G.3.B	Adding and Subtracting Fractions Through Decompositions		
4.NF.G.3.C	Adding and Subtracting Mixed Numbers		
4.NF.G.3.D	Adding and Subtracting Fractions in Word Problems		
4.NF.G.4.A	Multiplying Fractions		
4.NF.G.4.B	Multiplying Fractions by a Whole Number		
4.NF.G.4.C	Multiplying Fractions in Word Problems		
4.NF.H.5	10 to 100 Equivalent Fractions		
4.NF.H.6	Convert Fractions to Decimals		
4.NF.H.7	Compare Decimals		

Standard	Lesson	Score	Date of Completion
4.MD.I.1	Units of Measurement		
4.MD.I.2	Measurement Problems		
4.MD.I.3	Perimeter & Area		
4.MD.J.4	Representing and Interpreting Data		
4.MD.K.6	Angle Measurement		
4.MD.K.6	Measuring Turned Angles		
4.MD.K.6	Measuring and Sketching Angles		
4.MD.K.7	Adding and Subtracting Angle Measurements		
4.G.L.1	Points, Lines, Rays and Segments		
4.G.L.1	Angles		
4.G.L.2	Classifying Plane (2-D) Shapes		
4.G.L.3	Symmetry		

Lumos Learning
Step Up Your Skills

GRADE
4

WYOMING
WY-TOPP ENGLISH

Wyoming Test of Proficiency and Progress (WY-TOPP) Test Prep: Grade 4 English Language Arts Literacy (ELA) Practice Workbook and Full-length Online Assessments

REVISED EDITION

Workbook + Online Practice

Workbook Includes
- ✔ Worksheets to Practice Every Standard
- ✔ Answer Keys and Detailed Explanations

Online Program Includes
- ✔ 2 Full-Length Practice Tests
- ✔ Personalized Study Plan
- ✔ Progress Reports
- ✔ AI Tutor and More

100% aligned to the Wyoming Content and Performance Standards

Available
- At Leading book stores
- Online www.LumosLearning.com

www.ingramcontent.com/pod-product-compliance
Lightning Source LLC
Chambersburg PA
CBHW041113120626
46547CB00019B/2690